ADVENTURES UNDERGROUND

EXPLOSION OF FIRE-DAMP.

ADVENTURES UNDERGROUND

VOLCANOES, EARTHQUAKES, MINES, QUARRIES, CAVES AND PEAT BEDS

WITH NUMEROUS ILLUSTRATIONS

Fredonia Books
Amsterdam, The Netherlands

Adventures Underground:
Volcanoes, Earthquakes, Mines,
Quarries, Caves and Peat Beds

Compiled and Edited by
David Falkayn

ISBN: 1-58963-767-4

Reprinted from the 1897 edition

Fredonia Books
Amsterdam, the Netherlands
http://www.fredoniabooks.com

ADVENTURES
UNDERGROUND

CONTENTS

VOLCANOES AND EARTHQUAKES

Part I.—What they are

Part II.—Why they are.

MINES.

QUARRIES.

CAVES.

PEAT BEDS.

LIST OF ILLUSTRATIONS.

VOLCANOES AND EARTHQUAKES

PART I.

WHAT THEY ARE.

VOLCANOES AND EARTHQUAKES.

PART I.—WHAT THEY ARE.

CHAPTER I.

NEAR HOME.

THE first great earthquake of which any very distinct knowledge has reached us is that which occurred in the year 63 after our Saviour, which produced great destruction in the neighbourhood of Vesuvius, and shattered the cities of Pompeii and Herculaneum upon the Bay of Naples, though it did not destroy them.

This earthquake is chiefly remarkable as having been the forerunner and the warning (if that warning could have been understood) of the first eruption of Vesuvius on record, which followed sixteen years afterwards, in the year 79. Before that time none of the ancients had any notion of its being a volcano, though Pompeii itself is

paved with its lava. The crater was probably filled, or
at least the bottom occupied by a lake; and we read of
it as the stronghold of the rebel chief Spartacus, who,
when lured there by the Roman army, escaped with his
followers by clambering up the steep sides by the help of
the wild vines that festooned them.

The ground since the first earthquake in 63 had often
been shaken by slight shocks, when at length, in August,
79, they became more numerous and violent, and, on
the night preceding the eruption, so tremendous as to
threaten everything with destruction.

A morning of comparative repose succeeded, and the
terrified inhabitants of those devoted towns no doubt
breathed more freely, and hoped the worst was over;
when, about one o'clock in the afternoon, the Elder Pliny,
who was stationed in command of the Roman fleet at
Misenum, in full view of Vesuvius, beheld a huge black
cloud rising from the mountain, which, "rising slowly
always higher," at last spread out aloft like the head of
one of those picturesque flat-topped pines which form
such an ornament of the Italian landscape. The meaning
of such a phenomenon was to Pliny and to every one a
mystery. We know now too well what it imports, and
they were not long left in doubt. From that cloud descended stones, ashes, and pumice; and the cloud itself
lowered down upon the surrounding country, involving
land and sea in profound darkness, pierced by flashes of
fire more vivid than lightning. These, with the volumes
of ashes that began to encumber the soil, and which

covered the sea with floating pumice-stone, the constant heaving of the ground, and the sudden recoil of the sea, form a picture which is wonderfully well described by the Younger Pliny. His uncle, animated by an eager desire to know what was going on, and to afford aid to the

VESUVIUS IN ERUPTION.

inhabitants of the towns, made sail for the nearest point of the coast and landed; but was instantly enveloped in the dense sulphureous vapour that swept down from the mountain, and perished miserably.

It does not seem that any *lava* flowed on that occasion. Pompeii was buried under the ashes; Herculaneum by a torrent of mud, probably the contents of the crater, ejected at the first explosion.

This was most fortunate. We owe to it the preservation of some of the most wonderful remains of antiquity. For it is not yet much more than a century ago that, in digging a well at Portici, near Naples, the Theatre of Herculaneum was discovered, some sixty feet under ground; then houses, baths, statues, and, most interesting of all, a library, full of books, and those books still legible, and among them the writings of some ancient authors which had never before been met with, but which have now been read, copied, and published, while hundreds and hundreds, I am sorry to say, still remain unopened.

Pompeii was not buried so deep; the walls of some of the buildings appeared among the modern vineyards, and led to excavations, which were easy, the ashes being light and loose. And there you now may walk through the streets, enter the houses, and find the skeletons of their inmates, some in the very act of trying to escape. Nothing can be more strange and striking.

Since that time Vesuvius has been frequently but very irregularly in eruption. But I shall not occupy attention with a history of Vesuvius, but pass at once to the eruption of 1779, one of the most interesting on record, from the excellent account given of it by Sir William Hamilton, who was then resident at Naples as our

Minister, and watched it throughout with the eye of an artist as well as the scrutiny of a philosopher.

In 1767, there had been a considerable eruption, during which Pliny's account of the great pine-like, flat-topped, spreading mass of smoke had been superbly seen, extending over the Island of Capri, which is twenty-eight miles from Vesuvius. The showers of ashes, the lava currents, the lightnings, thunderings, and earthquakes were very dreadful; for from that time till the year 1779 Vesuvius never was quiet. In the spring of that year it began to pour out lava; and on one occasion, when Sir William Hamilton approached too near, the running stream was on the point of surrounding him, and the sulphureous vapour cut off his retreat, so that his only mode of escape was to walk across the lava, which, to his astonishment, and no doubt to his great joy, he found accompanied with no difficulty, and with no more inconvenience than what proceeded from the radiation of heat, on his legs and feet, from the scoriæ and cinders with which the external crust of the lava was loaded, and which in great measure intercepted and confined the glowing heat of the ignited mass below.

In such cases, and when cooled down to a certain point, the motion of the lava-stream is slow and creeping; rather rolling over itself than flowing like a river.

However, as time went on, this quiet mode of getting rid of its contents would no longer suffice, and the usual symptoms of more violent action——rumbling noises and explosions within the mountain, puffs of smoke from its crater, and jets of red-hot stones and ashes—continued till the end of July, when they increased to such a degree as to exhibit at night the most beautiful firework imaginable. The eruption came to its climax from the 5th to the 10th of August, on the former of which days, after the ejection of an enormous volume of white clouds, piled like bales of the whitest cotton, in a mass exceeding four times the height and size of the mountain itself, the lava began to overflow the rim of the crater, and stream in torrents down the steep slope of the cone.

"About nine o'clock," says Sir William Hamilton, on Sunday the 8th of August, "there was a loud report, which shook the houses at Portici and its neighbourhood to such a degree as to alarm the inhabitants and drive them out into the streets. Many windows were broken, and, as I have since seen, walls cracked by the concussion of the air from that explosion. . . . In one instant a fountain of liquid transparent fire began to rise, and gradually increasing, arrived at so amazing a height, as to strike every one who beheld it with the most awful astonishment. I shall scarcely be credited when I assure you that, to the best of my judgment, the height of this stupendous column of fire could not be less than three times that of Vesuvius itself, which, you know, rises perpendicularly near 3,700 feet above the level of the sea." (The height as measured in 1824 was 3,920 feet.)

"Puffs of smoke, as black as can possibly be imagined, succeeded one another hastily, and accompanied the red-hot, transparent, and liquid lava, interrupting its splendid brightness here and there by patches of the darkest hue. Within these puffs of smoke at the very moment of their emission from the crater, I could perceive a bright but pale electrical fire playing about in zigzag lines. The liquid lava, mixed with scoriæ and stones, after having mounted, I verily believe, at least 10,000 feet, falling perpendicularly on Vesuvius, covered its whole cone, part of that of Somma, and the valley between them. The falling matter being nearly as vivid and inflamed as that which was continually issuing fresh from the crater, formed with it one complete body of fire, which could not be less than two miles and a half in breadth, and of the extraordinary height above mentioned, casting a heat to the distance of at least six miles around it.

"The brushwood of the mountain of Somma was soon in a flame, which being of a different tint from the deep red of the matter thrown out from the volcano, and from the silvery blue of the electrical fire, still added to the contrast of this most extraordinary scene. After the column of fire had continued in full force for near half an hour, the eruption ceased at once, and Vesuvius remained sullen and silent."

To give an idea of the state of the inhabitants of the country when an explosion is going on, I will make one other extract :—

"The mountain of Somma, at the foot of which Ottaiano is situated, hides Vesuvius from its sight, so that till the eruption became considerable it was not visible to the inhabitants of Ottaiano. On Sunday night, when the noise increased, and the fire began to appear above the mountain of Somma, many of them flew to the churches, and others were preparing to quit the town, when a sudden violent report was heard, soon after which they found themselves involved in a thick cloud of smoke and minute ashes: a horrid clashing noise was heard in the air, and presently fell a deluge of stones and large scoriæ, some of which scoriæ were of the diameter of seven or eight feet, and must have weighed more than one hundred pounds before they were broken by their falls, as some of the fragments of them which I picked up in the streets still weighed upwards of sixty pounds.

"When these large vitrified masses either struck against each other in the air or fell on the ground, they broke in many pieces, and covered a large space around them with vivid sparks of fire, which communicated their heat to everything that was combustible.

"In an instant the town and country about it was on fire in many parts; for in the vineyards there were several straw-huts, which had been erected for the watchmen of the grapes, all of which were burnt. A great magazine of wood in the heart of the town was all in a blaze, and had there been much wind, the flames must have spread universally, and all the inhabitants would have infallibly been burnt in their houses, for it was impossible for them to stir out. Some who attempted it with pillows, tables, chairs, tops of wine-casks, &c., on their heads, were either knocked down or driven back to their close quarters, under arches and in the cellars of the houses.

AETNA IN ERUPTION

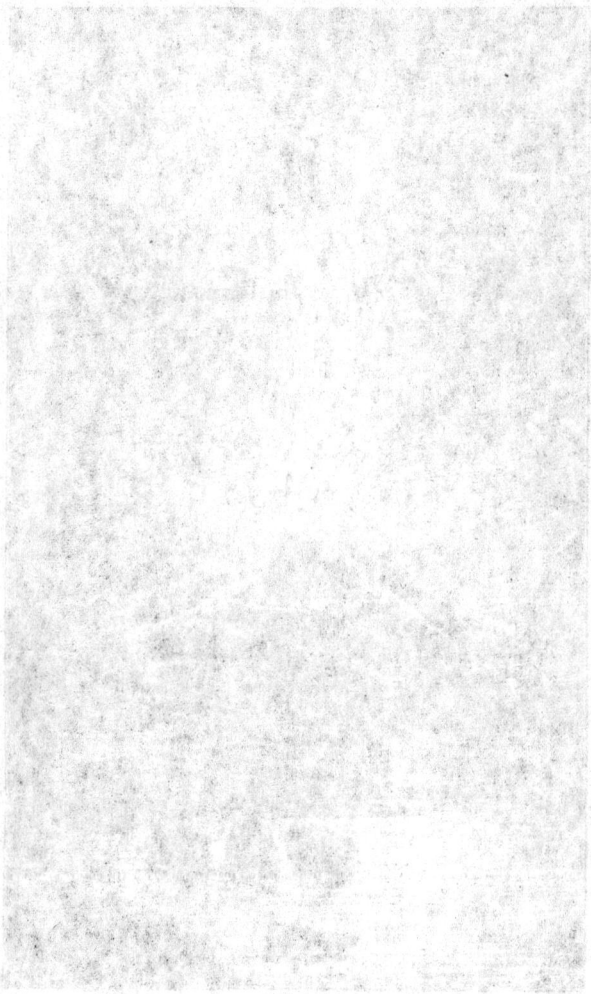

Many were wounded, but only two persons have died of the wounds they received from this dreadful volcanic shower. To add to the horror of the scene, incessant volcanic lightning was writhing about the black cloud that surrounded them, and the sulphureous smell and heat would scarcely allow them to draw their breath."

The next volcano I shall introduce is Ætna, the grandest of all our European volcanoes, 10,772 feet above the sea.

The scenery of Ætna is on the grandest scale. Ascending from Catania, you skirt the stream of lava which destroyed a large part of that city in 1669, and which ran into the sea, forming a jetty or breakwater that now gives Catania what it never had before, the advantage of a harbour. There it lies as hard, rugged, barren, and fresh-looking as if it had flowed but yesterday. In many places it is full of huge caverns, great air-bubbles, into which one may ride on horseback (at least large enough) and which communicate, in a succession of horrible vaults, where one might wander and lose one's self without hopes of escape.

Higher up, near Nicolosi, is the spot from which that lava flowed. It is marked by two volcanic cones, each of them a considerable mountain, called the Monti Rossi, rising three hundred feet above the slope of the hill, and which were thrown up on that occasion. Indeed, one of the most remarkable features of Ætna is that of its flanks bristling over with innumerable smaller volcanoes. For the height is so great that the lava now scarcely ever rises to the top of the crater, for before that its immense weight breaks through at the sides.

From the summit of Ætna extends a view of extraordinary magnificence. The whole of Sicily lies at your feet, and far beyond it are seen a string of lesser volcanoes, the Lipari Islands, between Sicily and the Italian coast, one of which, Stromboli, is always in eruption, unceasingly throwing up ashes, smoke, and liquid fire.

But I must not linger on the summit of Ætna. We will now take a flight thence, all across Europe, to Iceland—a wonderful land of frost and fire. It is full of volcanoes, one of which, Hecla, has been twenty-two times in eruption within the last eight hundred years.

Besides Hecla, there are five others, from which in the same period twenty eruptions have burst forth, making about one every twenty years. The most formidable of these was that which happened in 1783, a year also memorable as that of the terrible earthquake· in Calabria.

In May of that year, a bluish fog was observed over the mountain called Skaptar Jokul, and the neighbourhood was shaken by earthquakes. After a while a great pillar of smoke was observed to ascend from it, which darkened the whole surrounding district, and which descended in a whirlwind of ashes. On the 10th of May, innumerable fountains of fire were seen shooting up through the ice and snow which covered the mountain; and the principal river, called the Skapta, after rolling

HECLA.

down a flood of foul and poisonous water, disappeared.

Two days after, a torrent of lava poured down into the bed which the river had deserted. The river had run in a deep ravine, six hundred feet deep and two hundred broad. This the lava entirely filled; and not only so, but it overflowed the surrounding country, and ran into a great lake, from which it instantly expelled the water in an explosion of steam. When the lake was fairly filled, the lava again overflowed and divided into two streams, one of which covered some ancient lava fields; the other re-entered the bed of the Skapta lower down, and presented the astounding sight of a cataract of liquid fire pouring over what was formerly the waterfall of Stapafoss.

This was the greatest eruption on record in Europe. It lasted in its violence till the end of August, and closed with a violent earthquake; but for nearly the whole year a canopy of cinder-laden cloud hung over the island; the Faroe Islands, nay, even Shetland and the Orkneys, were deluged with the ashes, and volcanic dust and a preternatural smoke which obscured the sun, covered all Europe as far as the Alps, over which it could not rise.

I have little doubt that the great Fire-ball of August 18, 1788, which traversed all England and the Continent, from the North Sea to Rome, by far the greatest ever known (for it was more than half a mile in diameter), was somehow connected with the electric excitement of the upper atmosphere produced by this enormous discharge of smoke and ashes.

c

The destruction of life in Iceland was frightful : 9,000 men, 11,000 cattle, 28,000 horses, and 190,000 sheep perished ; mostly by suffocation. The lava ejected has been computed to have amounted in volume to more than twenty cubic miles.

VOLCANOES AND EARTHQUAKES.

CHAPTER II.

FAR AWAY.

WE shall now proceed to still more remote regions, and describe, in as few words as may be, two immense eruptions—one in Mexico in the year 1759, the other in the island of Sumbawa, in the Eastern Archipelago, in 1815.

I ought to mention, by way of preliminary, that almost the whole line of coast of South and Central America, from Mexico southwards as far as Valparaiso—that is to say, nearly the whole chain of the Andes—is one mass of volcanoes. In Mexico and Central America there are two-and-twenty, and in Quito, Peru, and Chili, six-and-twenty more, in activity; and nearly as many more apparently extinct ones, any one of which may at any moment break out afresh. This does not prevent the country from being inhabited, fertile, and well cultivated.

Well, in a district of Mexico celebrated for the growth of the finest cotton, between two streams called Cuitimba and San Pedro, which furnished water for irrigation, lay the farm and homestead of Don Pedro de Jurullo, one of the richest and most fertile properties in that country. He was a thriving man, and lived in comfort as a large proprietor, little expecting the mischief that was to befall him. In June, 1759, however, a subterraneous noise was heard in this peaceful region.

Hollow sounds of the most alarming nature were succeeded by frequent earthquakes, succeeding one another for fifty or sixty days; but they died away, and in the beginning of September everything seemed to have returned to its usual state of tranquillity. Suddenly, on the night of the 28th of September, the horrible noises recommenced. All the inhabitants fled in terror; and the whole tract of ground, from three to four square miles in extent, rose up in the form of a bladder to a height of upwards of five hundred feet !

Flames broke forth over a surface of more than half a square league, and through a thick cloud of ashes illuminated by this ghastly light, the refugees, who had ascended a mountain at some distance, could see the ground as if softened by the heat, and swelling and sinking like an agitated sea. Vast rents opened in the earth, into which the two rivers I mentioned precipitated themselves, but so far from quenching the fires, only seemed to make them more furious. Finally, the whole plain became covered with an immense torrent of boiling mud, out of which sprang thousands of little volcanic cones called Hornitos, or ovens.

But the most astonishing part of the whole was the opening of a chasm vomiting out fire, and red-hot stones, and ashes, which accumulated so as to form a range of six large mountain masses, one of which is upwards of sixteen hundred feet in height above the old level, and which is now known as the volcano of Jurullo. It is continually burning; and for a whole year continued to throw up an immense quantity of ashes, lava, and fragments of rock.

The roofs of houses at the town or village of Quere-taro, upwards of a hundred and forty miles distant, were covered with the ashes.

The two rivers have again appeared, issuing at some distance from among the hornitos, but no longer as sources of wealth and fertility, for they are scalding hot, or at least were so when Baron Humboldt visited them several years after the event. The ground even then retained a violent heat, and the hornitos were pouring forth columns of steam twenty or thirty feet high, with a rumbling noise like that of a steam-boiler.

The island of Sumbawa is one of that curious line of islands which links on Australia to the south-eastern corner of Asia. On this island is the volcano of Tomboro, which broke out into eruption on the 5th of April in 1815; and I can hardly do better than quote the account of it in Sir Stamford Raffles's own words.

"Almost every one," says this writer, "is acquainted with the intermitting convulsions of Ætna and Vesuvius;

but the most extraordinary of them can bear no comparison, in point of duration and force, with that of Mount Tomboro, in the island of Sumbawa.

"This eruption extended perceptible evidences of its existence over the whole of the Molucca Islands, over Java, a considerable portion of the Celebes, Sumatra, and Borneo, to a circumference of 1,000 statute miles from its centre" (*i.e.* to 1,000 miles' *distance*), "by tremulous motions and the report of explosions. In a short time the whole mountain near the Sang'ir appeared like a body of liquid fire, extending itself in every direction. The fire and columns of flame continued to rage with unabated fury, until the darkness, caused by the quantity of falling matter, obscured it at about eight P.M. Stones at this time fell very thick at Sang'ir, some of them as large as two fists, but generally not larger than walnuts.

"Between nine and ten P.M., ashes began to fall, and soon after a violent whirlwind ensued, which blew down nearly every house of Sang'ir, carrying the roofs and light parts away with them.

"In the port of Sang'ir, adjoining Sumbawa, its effects were much more violent, tearing up by the roots the largest trees, and carrying them into the air, together with men, horses, cattle, and whatsoever came within its influence. This will account for the immense number of floating trees seen at sea.

"The sea rose nearly twelve feet higher than it had ever been known to do before, and completely spoiled the only small spots of rice land in Sang'ir, sweeping away houses and everything within its reach. The whirlwind lasted about an hour.

"No explosions were heard till the whirlwind had ceased at about eleven A.M. From midnight till the evening of the 11th they continued without intermission; after that time their violence moderated, and they were heard only at intervals; but the explosions did not cease entirely until the 15th of July.

"Of all the villages round Tomboro, Tempo, containing about forty inhabitants, is the only one remaining. In Pekaté no vestige of a house is left; twenty-six of the people, who were at Sumbawa at the time, are the whole of the population who have escaped. From the best inquiries, there were certainly not fewer than 12,000 individuals in Tomboro and Pekaté at the time of the eruption, of whom five or six survive. The trees and herbage of every description, along the whole of the north and west of the peninsula, have been completely destroyed, with the exception of a high point of land near the spot where the village of Tomboro stood. At Sang'ir, it is added, the famine occasioned by this event was so extreme, that one of the rajah's own daughters died of starvation."

I have seen it computed that the quantity of ashes and lava vomited forth in this awful eruption would have formed three mountains the size of Mont Blanc, the highest of the Alps; and if spread over the surface of Germany, would have covered the whole of it two feet deep! The ashes did actually cover the whole island of Tombock, more than a hundred miles distant, to that depth, and 44,000 persons there perished by starvation, from the total destruction of all vegetation.

The mountain Kilauea, in the island of Hawaii, one of the Sandwich Isles, exhibits the remarkable phenomenon of a lake of molten and very liquid lava *always* filling the bottom of the crater, and always in a state of terrific ebullition, rolling to and fro its fiery surge and flaming billows—yet with this it is content, for it would seem that at least for a long time past there has been no violent outbreak so as to make what is generally understood by a volcanic eruption.

Volcanic eruptions are almost always preceded by earthquakes, by which the beds of rock, that overlie and keep down the struggling powers beneath, are dislocated and cracked, till at last they give way, and the strain is immediately relieved. It is chiefly when this does not happen, when the force below is sufficient to heave up and shake the earth, but not to burst open the crust, and give vent to the lava and gases, that the most destructive effects are produced.

The great earthquake of November 1, 1755, which destroyed Lisbon, was an instance of this kind, and was one of the greatest, if not the very greatest, on record; for the concussion extended over all Spain and Portugal—indeed over all Europe, and even into Scotland—over North Africa, where in one town in Morocco 8,000 or 10,000 people perished. Nay, its effects extended even across the Atlantic to Madeira, where it was very violent, and to the West Indies.

LISBON DURING THE EARTHQUAKE.

The most striking feature about this earthquake was its extreme suddenness. All was going on quite as usual in Lisbon the morning of that memorable day, the weather fine and clear, and nothing whatever to give the population of that great capital the least suspicion of mischief. All at once, at twenty minutes before ten A.M., a noise was heard like the rumbling of carriages under ground; it increased rapidly and became a succession of deafening explosions like the loudest cannon. Then a shock, which, as described by one writing from the spot, seemed to last but the tenth part of a minute, and down came tumbling palaces, churches, theatres, and every large public edifice, and about a third or a fourth part of the dwelling-houses. More shocks followed in rapid succession, and in six minutes from the commencement 60,000 persons were crushed in the ruins!

Here are the simple but expressive words of one J. Latham, who writes to his uncle in London :—

" I was on the river with one of my customers going to a village three miles off. Presently the boat made a noise as if on the shore or landing, though then in the middle of the water. I asked my companion if he knew what was the matter. He stared at me, and looking at Lisbon, we saw the houses falling, which made him say, ' God bless us, it is an earthquake !' About four or five minutes after the boat made a noise as before, and we saw the houses tumble down on both sides of the river."

They then landed and made for a hill; thence they beheld the sea (which had at first receded and laid a great tract dry) come rolling in, in a vast mountain wave fifty or sixty feet high, on the land, and sweeping all before it.   Three thousand people had taken refuge on a new stone quay or jetty just completed at great expense. In an instant it was turned topsy-turvy; and the whole quay, and every person on it, with all the vessels moored to it, disappeared, and not a vestige of them ever appeared again.   Where that quay stood was afterwards found a depth of a hundred fathoms (six hundred feet) water.

It happened to be a religious festival, and most of the population were assembled in the churches, which fell and crushed them.

That no horror might be wanting, fires broke out in innumerable houses where the wood-work had fallen on the fires, and much that the earthquake had spared was destroyed by fire.   And then too broke forth that worst of all scourges, a lawless ruffian-like mob who plundered, burned, and murdered in the midst of all that desolation and horror.

The huge wave I have spoken of swept the whole coast of Spain and Portugal.   Its swell and fall was ten or twelve feet at Madeira.   It swept quite across the Atlantic, and broke on the shores of the West Indies. Every lake and firth in England and Scotland was dashed for a moment out of its bed, the water not partaking of the sudden *shove* given to the land, just as when you splash a flat saucerful of water, the water dashes over on the side *from* which the shock is given.

One of the most curious incidents in this earthquake

was its effect on ships far out at sea, which would lead us to suppose that the immediate impulse was in the nature of a violent blow or thrust upwards under the bed of the ocean.

Thus it is recorded that this upward shock was so sudden and violent on a ship, at that time forty leagues from Cape St. Vincent, that the sailors on deck were tossed up into the air to a height of eighteen inches. A British ship eleven miles from land near the Philippine Islands, in 1796, was struck upwards with such force as to unship and split up the main-mast.

The same kind of upward bounding movement took place at Riobamba in Quito in the great earthquake of February 4, 1797, which was connected with an eruption of the volcano of Tunguragua. That earthquake extended in its greatest intensity over an oval space of a hundred and twenty miles from south to north, and sixty from east to west, within which space every town and village was levelled with the ground; but the total extent of surface shaken was upwards of five hundred miles in one direction (from Puna to Popayan), and four hundred in the other. Quero, Riobamba, and several other towns, were buried under fallen mountains, and in a very few minutes 30,000 persons were destroyed. At Riobamba, however, after the earthquake, a great number of corpses were found to have been tossed across a river, and scattered over the side of a sloping hill on the other side.

The frequency of these South American earthquakes is not more extraordinary than the duration of the shocks.

Humboldt relates that on one occasion, when travelling on mule-back with his companion Bonpland, they were obliged to dismount in a dense forest, and throw themselves on the ground, the earth being shaken uninterruptedly for upwards of a quarter of an hour with such violence that they could not keep their legs.

One of the most circumstantially described earthquakes on record is that which happened in Calabria on the 5th of February, 1783,—I should say began then, for it may be said to have lasted four years. In the year 1788, for instance, 949 shocks took place, of which 501 were great ones, and in 1784, 151 shocks were felt, 98 of which were violent.

The centre of action seemed to be under the towns of Monteleone and Oppido. In a circle twenty-two miles in radius round Oppido every town and village was destroyed within two minutes by the first shock, and within one of seventy miles' radius all were seriously shaken and much damage done. The whole of Calabria was affected, and even across the sea Messina was shaken, and a great part of Sicily.

There is no end of the capricious and out-of-the-way accidents and movements recorded in this Calabrian earthquake.

The ground undulated like a ship at sea, people became actually sea-sick, and to give an idea of the undulation (just as it happens at sea), the scud of the clouds before the wind seemed to be fitfully arrested during the pitching movement when it took place in the same direction, and to redouble its speed in the reverse movement. At Oppido many houses were swallowed up bodily. Loose objects were tossed up several yards into the air. The flagstones in some places were found after a severe shock all turned

FISSURE, DURING EARTHQUAKE.

bottom upwards. Great fissures opened in the earth, and at Terra Nova a mass of rock two hundred feet high and four hundred in diameter travelled four miles down a ravine. All landmarks were removed, and the land itself, in some instances with trees and hedges growing on it, carried bodily away and set down in another place.

Altogether about 40,000 people perished by the earth. quakes, and some 20,000 more of the epidemic diseases produced by want and the effluvia of the dead bodies.

Volcanoes occasionally break forth at the bottom of the sea, and, when this is the case, the result is usually the

BARREN ISLAND.

production of a new island. This, in many cases, disappears soon after its formation, being composed of loose and incoherent materials, which easily yield to the destructive power of the waves.

Such was the case with the island of Sabrina, thrown up, in 1811, off St. Michael's, in the Azores, which disappeared almost as soon as formed, and in that of Pantellaria, on the Sicilian coast, which resisted longer, but was gradually washed into a shoal, and at length has, we believe, completely disappeared. In numerous other instances, the cones of cinders and scoriæ, once raised, have become compacted and bound together by the effusion of lava, hardening into solid stone, and thus, becoming habitual volcanic vents, they continue to increase

TENERIFFE.

in height and diameter, and assume the importance of permanent volcanic islands.

Such has been, doubtless, the history of those numerous insular volcanoes which dot the ocean in so many parts of the world, such as Teneriffe, the Azores, Ascension, St. Helena, Tristan d'Aumbra, Barren Island, &c. In

D

some cases the process has been witnessed from its commencement, as in that of two islands which arose in the
Aleutian group, connecting Kamtschatka with North
America, the one in 1796, the other in 1814, and which
both attained the elevation of three thousand feet.

Besides these evident instances of eruptive action, there
is every reason to believe that enormous floods of lava
have been, at various remote periods in the earth's
history, poured forth at the bottom of seas so deep as to
repress, by the mere weight of water, all outbreak of
steam, gas, or ashes ; and reposing perhaps for ages in a
liquid state, protected from the cooling action of the
water on their upper surface by a thick crust of congealed
stony matter, to have assumed a perfect level, and, at
length, by slow cooling, taken on that peculiar columnar
structure which we see produced in miniature in starch
by the contraction or shrinkage, and consequent splitting,
of the material in drying; and resulting in those picturesque and singular landscape features called basaltic
colonnades, when brought up to day by sudden or gradual
upheaval, and broken into cliffs and terraces by the action
of waves, torrents, or weather.

Those grand specimens of such colonnades which
Britain possesses in the Giant's Causeway of Antrim, and
the Cave of Fingal in Staffa, for instance, are no doubt
extreme outstanding portions of such a vast submarine
lava-flood which at some inconceivably remote epoch
occupied the whole intermediate space, affording the same
kind of evidence of a former connection of the coasts of

Scotland and Ireland as do the opposing chalk cliffs of Dover and Boulogne of the ancient connection of France with Britain. Here and there a small basaltic island, such as that of Rathlin, remains to attest this former continuity, and to recall to the contemplative mind that sublime antagonism between sudden violence and perse vering effort. which the study of geology impresses in every form of repetition.

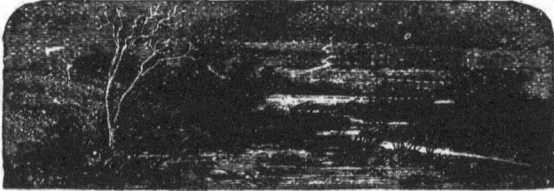

VOLCANOES AND EARTHQUAKES.

CHAPTER III.

THEIR CAUSE.

WHAT is the cause of an earthquake?

The question seems a profound one, and yet to find its true answer let us try the very simplest cause of which we can think. That is the wise and scientific plan.

Now, whatever makes these earthquakes must of course be enormously strong. That is certain. And what is the strongest thing you know of in the world? Think. . . . . . .

Gunpowder?

Well, gunpowder is strong sometimes, but not always. You may carry it in a flask, or in your hand, and then it is weak enough. It only becomes strong by being turned into gas and steam. But steam is always strong. And if you look at a railway engine, still more if you had ever seen—which God forbid you should—a boiler explosion, you would agree with me, that the strongest thing we know of in the world is steam.

Now I think that we can explain almost, if not quite, all that we know about earthquakes, if we believe that on the whole they are caused by steam and other gases expanding, that is, spreading out, with wonderful quickness and strength. Of course there must be something to make them expand, and that is *heat*. But we will not talk of that yet.

You know why the lid of a kettle rattles. The expanding steam escapes in little jets, and so causes a *lid-quake*. Now suppose that there was steam under the earth trying to escape, and the earth in one place was loose and yet hard, as the lid of the kettle is loose and yet hard, with cracks in it, it may be, like the crack between the edge of the lid and the edge of the kettle itself; might not the steam try to escape through the cracks, and rattle the surface of the earth, and so cause an *earth-quake?*

So the steam would escape generally easily, and would only make a passing rattle, like the earthquake of which the famous jester Charles Selwyn said, that it was quite a young one, so tame that you might have stroked it; like that which I myself once felt in the Pyrenees, which gave me very solemn thoughts after a while, though at first I did nothing but laugh at it; and I will tell you why.

I was travelling in the Pyrenees, and I came one
evening to the loveliest spot—a glen, or rather a vast

GLEN IN THE PYRENEES.

crack, in the mountains, so narrow that there was no
room for anything at the bottom of it, save a torrent
roaring between walls of polished rock.   High above the

torrent the road was cut out among the cliffs, and above the road rose more cliffs, with great black cavern mouths, hundreds of feet above our heads, out of each of which poured in foaming waterfalls streams large enough to turn a mill, and above them mountains piled on mountains, all covered with woods of box, which smelt rich and hot and musky in the warm summer air. Among the box-trees and fallen boulders grew hepaticas, blue and white and red, such as you see in the garden; and little stars of gentian, more azure than the azure sky. But out of the box-woods above rose giant silver firs, clothing the cliffs and glens with tall black spires, till they stood out at last in a jagged saw-edge against the purple evening sky, along the mountain ranges, thousands of feet aloft; and beyond them again, at the head of the valley, rose vast cones of virgin snow, miles away in reality, but looking so brilliant and so near that one fancied at the first moment that one could have touched them with one's hand. Snow-white they stood, the glorious things, seven thousand feet into the air; and I watched their beautiful white sides turn rose-colour in the evening sun, and when he set fade into dull cold grey, till the bright moon came out to light them up once more.

When I was tired of wondering and admiring, I went into bed; and there I had a dream—such a dream as Alice had when she went into Wonderland—such a dream as I dare say you may have had ere now. Some noise or stir puts into your fancy as you sleep a whole long dream to account for it; and yet that dream, which seems to you to be hours long, has not taken up a second of time, for the very same noise which begins the dream wakes you at the end of it: and so it was with me.

I dreamed that some English people had come to the hotel where I was, and were sleeping in the room underneath me; and that they had quarrelled and fought, and broke their bed down with a tremendous crash, and that I must get up and stop the fight; and at that moment I awoke, and heard coming up the valley from the north such a roar as I never heard before or since, as if a hundred railway trains were rolling underground; and just as it passed under my bed there was a tremendous thump, and I jumped out of bed quicker than I ever did in my life, and heard the roaring sound die away as it rolled up the valley towards the peaks of snow. Still I had in my head this notion of the Englishmen fighting in the room below. But then I recollected that no Englishmen had come in the night before, and that I had been in the room below, and that there was no bed in it. Then I opened my window—a woman screamed, a dog barked, some cocks and hens cackled in a very disturbed humour, and then I could hear nothing but the roaring of the torrent a hundred feet below.

Next morning I inquired whether any one had heard a noise. No, nobody had heard anything. And the driver who had brought me up the valley only winked, but did not choose to speak. At last at breakfast I asked the pretty little maid who waited what was the meaning of the noise I heard in the night, and she answered, to my intense amusement, "Ah! bah! ce n'était qu'un tremblement de terre; il y en a ici toutes les six semaines."

And now the secret was out. The little maid, I found, came from the lowland far away, and did not mind telling the truth; but the good people of the place were afraid to let out that they had earthquakes every six weeks, for fear of frightening visitors away : and because they were really very good people, and very kind to me, I shall not tell you what the name of the place is.

Of course, after that I could do no less than ask how earthquakes should occur in that particular place, hundreds of miles away from any burning mountain ?

As I had come up the valley I had seen that the cliffs were all beautiful grey limestone marble; but just at this place they were replaced by granite, such as you may see in the material of London Bridge or in the hills at Aberdeen. I do not mean that the limestone changed to granite, but that the granite had risen up out of the bottom of the valley, and had carried the limestone (I suppose) up on its back hundreds of feet into the air. But I knew that beside these cold springs there were hot springs also, full of curious chemical salts, just below the very house where I was in. And when I went to look at them, I found that they came out of the rock just where the limestone and the granite joined. "Ah, ha!" I said, "now I think I have the answer. The lid of one of nature's great steam boilers is rather shaky and cracked

just here, because the granite has broken and torn the
limestone as it lifted it up ; and here is the hot water out
of the boiler actually oozing out of the crack ; and the
earthquake I heard last night was simply the steam
rumbling and thumping inside, and trying to get
out."

And then I fell into a more serious mood. I said
to myself, "If that steam had been a little, only a little
stronger, or the rock above it only a little lighter or
weaker, the village might have been shaken to the
ground, the rocks hurled into the torrent, jets of steam
and of hot water, mixed, it may be, with deadly gases,
have roared out of the riven ground ; that might have
happened here, in short, which has happened and
happens still in a hundred places in the world, whenever
the rocks are too weak to stand the pressure of the
steam below, and the solid earth bursts, as an engine-
boiler bursts when the steam within is too strong."

Most strange, but most terrible also, are the tricks
which this underground steam plays. It will make the
ground, which seems to us so hard and firm, roll and
rock in waves, till people are sea-sick, as on board a
ship ; and that rocking motion (which is the most
common) will often, when it is but slight, set the bells
ringing in the steeples, or make the furniture and things
on shelves jump about quaintly enough. It will make
trees bend to and fro, as if a wind was blowing through
them ; open doors suddenly, and shut them again with
a slam ; make the timbers of the floors and roofs creak,

as they do in a ship at sea; or give men such frights
as one of the dock-keepers at Liverpool got, in the
earthquake of 1863, when his watchbox rocked so, that
he thought some one was going to pitch him over into
the dock.

VOLCANOES AND EARTHQUAKES.

CHAPTER IV.

THEIR CAUSE.—(*Continued.*)

BUT these are only little hints and warnings of what it can do. When it is strong enough, it will rock down houses and churches into heaps of ruins, or, if it leaves them standing, crack them from top to bottom, so that they must be pulled down and rebuilt.

You see this picture of the ruins of Arica, in Peru, which was destroyed by an earthquake in 1868; and from it you can guess well enough for yourself what a town looks like which has been ruined by an earthquake. Of the misery and the horror which follow such a ruin I will not talk to you now. But the strangeness of some of the tricks which the earthquake shocks play is hardly to be explained, even by scientific men.

Sometimes, it would seem, the force runs round,

making the solid ground eddy, as water eddies in a brook. For it will make straight rows of trees crooked; it will twist whole walls round—or rather the ground on which the walls stand—without throwing them down; it will shift the stones of a pillar one on the other sideways, as if a giant had been trying to spin it like a teetotum, and so screwed it half in pieces. There is a story told by a wise man, who saw the place himself, of the whole furniture of one house being hurled away by the earthquake, and buried under the ruins of another house, and of things carried hundreds of yards off, so that the neighbours went to law to settle who was the true owner of them.

Sometimes, again, the shock seems to come neither horizontally in waves nor circularly in eddies, but vertically, that is, straight up from below; and then things— and people, alas! sometimes—are thrown up off the earth high into the air, just as things spring up off the table if you strike it smartly enough underneath. By that same law (for there is a law for every sort of motion) it is that the earthquake shock sometimes hurls great rocks off a cliff into the valley below. The shock runs through the mountain till it comes to the cliff at the end of it; and then the face of the cliff, if it be at all loose, flies off into the air.

You may see the very same thing happen, if you will put marbles or billiard-balls in a row touching each other, and strike the one nearest you smartly in the line of the row. All the balls stand still except the last one, and that flies off. The shock, like the earthquake shock, has run through them all, but only the end one, which had nothing beyond it but soft air, has been moved; and if you learn mathematics, you will know the law of motion according to which that happens, and learn to apply what the billiard-balls have taught you to explain the wonders of an earthquake. For in this case, as in so many more, you must watch little and common things to find out how nature works in great and rare ones. That is why Solomon says that " a fool's eyes are in the ends of the earth," because he is always looking out for strange things which he has not seen, and which he could not understand if he saw, instead of looking at the petty commonplace matters which are about his feet all day long, and getting from them sound knowledge, and the art of getting more sound knowledge still.

Another terrible destruction which the earthquake brings, when it is close to the seaside, is the wash of a great sea-wave. The sea moans, and sinks back, leaving the shore dry; and then comes in from the offing a mighty wall of water, as high as, or higher than, many a tall house; sweeps far inland, washing away quays and houses, and carrying great ships in with it; and then sweeps back again, leaving the ships high and dry.

Now, how is that wave made? Let us think. Perhaps in many ways. But two of them I will tell you as simply as I can, because they seem the most likely, and probably the most common.

Suppose, as the earthquake shock ran on, making the

ARICA, AFTER THE EARTHQUAKE

earth under the sea heave and fall in long earth-waves, the sea-bottom sank down. Then the water on it would sink down too, and leave the shore dry, till the sea-bottom rose again, and hurled the water up again against the land. This is one way of explaining it, and it may be true. For certain it is, that earthquakes do move the bottom of the sea; and certain, too, that they move the water of the sea also, and with tremendous force. For ships at sea during an earthquake feel such a blow from it (though it does them no harm) that the sailors often rush upon deck, fancying that they have struck upon a rock; and the force which could give a ship, floating in water, such a blow as that, would be strong enough to hurl thousands of tons of water up the beach, and on to the land.

But there is another way of accounting for this great sea-wave, which I fancy comes true sometimes.

Suppose you put an empty india-rubber ball into water, and then blew into it through a pipe. Of course, you know, as the ball filled, the upper side of it would rise out of the water. Now, suppose there were a party of little ants moving about upon that ball, and fancying it a great island, or perhaps the whole world—what would they think of the ball's filling and growing bigger?

If they could see the sides of the basin or tub in which the ball was, and were sure that they did not move, then they would soon judge by them that they themselves were moving, and that the ball was rising out of the

E

water. But if the ants were so short-sighted that they could not see the sides of the basin, they would be apt to make a mistake, because they would then be like men on an island out of sight of any other land. Then it would be impossible further to tell whether they were moving up or whether the water was moving down—whether their ball was rising out of the water or the water was sinking away from the ball. They would probably say, " The water is sinking, and leaving the ball dry."

Do you understand that? Then think what would happen if you pricked a hole in the ball. The air inside would come hissing out, and the ball would sink again into the water. But the ants would probably fancy the very opposite. Their little heads would be full of the notion that the ball was solid and could not move, just as our heads are full of the notion that the earth is solid and cannot move; and they would say, "Ah! here is the water rising again."

Just so, I believe, when the sea seems to ebb away during the earthquake, the land is really being raised out of the sea, hundreds of miles of coast, perhaps, or a whole island, at once, by the force of the steam and gas imprisoned under the ground. That steam stretches and strains the solid rocks below till they can bear no more, and snap and crack with frightful roar and clang; then out of holes and chasms in the ground rush steam, gases —often foul and poisonous ones—hot water, mud, flame, strange stones—all signs that the great boiler down below has burst at last

Then the strain is eased. The earth sinks together again, as the ball did when it was pricked, and sinks lower, perhaps, than it was before; and back rushes the sea, which the earth had thrust away while it rose, and sweeps in, destroying all before it.

Of course, there is a great deal more to be said about all this: but I have no time to tell you now. You will

VOLCANO AT SEA.

read it, I hope, for yourselves when you grow up, in the writings of far wiser men than I. Or perhaps you may feel for yourselves in foreign lands the actual shock of a great earthquake, or see its work fresh done around you. And if ever that happens, and you be preserved during the danger, you will learn for yourself, I trust, more about earthquakes than I can teach you, if you will only

bear in mind the simple general rules for understanding
the "how" of them which I have given you here.

But you do not seem satisfied yet? What is it that
you want to know?

Oh! There have been earthquakes here in England;
and you wonder will there ever be earthquakes in
England which will throw houses down and bury people
in the ruins?

I think you may set your heart at rest upon that
point. As far as the history of England goes back,
which is more than a thousand years, there is no account
of any earthquake which has done any serious damage,
or killed, I believe, a single human being.

The little earthquakes which are sometimes felt in
England run generally up one line of country, from
Devonshire through Wales, and up the Severn valley into
Cheshire and Lancashire, and the south-west of Scotland; and they are felt more smartly there, I believe,
because the rocks are harder there than here, and more
tossed about by earthquakes which happened ages and
ages ago, long before man lived on the earth. The work
of these earthquakes is seen in the tilting and twisting of
the layers of rock, and in the cracks (*faults*, as they are
called) which run through them in different directions,
which were made long, long ago. But even in the rocky
parts of England the earthquake force seems to have all
but died out. Perhaps the crust of the earth has become
too thick and solid there to be much shaken by the gases
and steam below. In the eastern part of England,

meanwhile, there is but little chance that an earthquake
will ever do much harm, because the ground there, for
thousands of feet down, is not hard and rocky, but soft—
sands, clays, chalk, and sands again; clays, soft limestones, and clays again—which all act as buffers to

EFFECT OF EARTHQUAKE ON ROCKS.

deaden the earthquake shocks, and deaden, too, the
earthquake noise.

And how?

Put your ear to one end of a soft bolster, and let some
one hit the other end. You will hear hardly any noise,
and will not feel the blow at all. Put your ear to one
end of a hard piece of wood, and let some one hit the
other. You will hear a smart tap, and perhaps feel a
smart tap, too. When you are older, and learn the laws

of sound, and of motion among the particles of bodies, you will know why. Meanwhile you may comfort yourself with the thought that there has been prepared for this good people of Britain a safe soft bed—not that they may lie and sleep on it, but work and till, plant and build and manufacture, and thrive in peace and comfort, we will trust and pray, for many a hundred years to come.

All that the steam inside the earth is likely to do to us is to raise parts of this island out of the sea so slowly, probably, that no man can tell whether they are rising or not. Or again, the steam-power may be even now dying out under our island, and letting parts of it sink slowly into the sea, as some think that the fens in Norfolk and Cambridgeshire are sinking now. Slowly but surely the whole world is changing around us. Ages ago Hartford Bridge Flat, for instance, was the gravelly bottom of a sea. Then the steam-power underground raised it up slowly, through long ages, till it became dry land. And ages hence, perhaps, it will have become a sea-bottom once more. Washed slowly away by the rain, or sunk by the dying out of the steam-power underground, it will go down again to the place from whence it came. Seas will roll where we stand now, and new lands will rise where seas now roll. For all things on this earth, from the tiniest flower to the tallest mountain, change, and change all day long. Every atom of matter moves perpetually, and nothing " continues in one stay."

The seeming solid earth on which you stand is but a

heaving bubble, bursting ever and anon in this place and in that. Only above all, and through all, and with all, is One who does not move nor change, but is the same yesterday, to-day, and for ever. And on Him, and not on this bubble of an earth, do you and I, and all mankind, depend.

VOLCANOES AND EARTHQUAKES.

CHAPTER V.

THEIR CONNECTION.

WHEREVER there are volcanoes there will be earthquakes. You may now and then have earthquakes without volcanoes; but seldom or never volcanoes without earthquakes.

How does that comes to pass? Does a volcano make earthquakes?

No; we may rather say, that earthquakes are trying to make volcanoes. For volcanoes are the holes which the steam underground has burst open, that it may escape into the air above. They are the chimneys of the great blast-furnaces underground, in which is pounded and melted up the old rocks, to make them into new ones, and spread them out over the land above.

And are there many volcanoes in the world? You
have read of Vesuvius, in Italy; and Ætna, in Sicily;
and Hecla, in Iceland. And you have heard, too, of
Kilauea, in the Sandwich Islands, and of Pele's Hair—
the yellow threads of lava, like fine spun glass, which
are blown from off its pools of fire, and which the
Sandwich Islanders believed to be the hair of a goddess
who lived in the crater. And you have read, too, I
hope, in Miss Yonge's "Book of Golden Deeds," the
noble story of the Christian chieftainess who, in order
to persuade her subjects to become Christians also, went
down into the crater and defied the goddess of the volcano, and came back unhurt and triumphant.

But if you look at Keith Johnston's Physical Atlas,
and look at the map which is called "Phenomena of
Volcanic Action," you will see that there are many,
many more.

You will see in it many red dots, which mark the
volcanoes which are still burning; and black dots, which
mark those which have been burning at some time or
other, not very long ago, scattered about the world.

Sometimes they are single, like the red dot at Otaheite,
or at Easter Island in the Pacific. Sometimes they are
in groups, or clusters, like the cluster at the Sandwich
Islands, or in the Friendly Islands, or in New Zealand.
And if we look in the Atlantic, we shall see four clusters;
one in poor half-destroyed Iceland, in the far north, one
in the Azores, one in the Canaries, and one in the Cape
de Verds. And there is one dot in those Canaries which

we must not overlook, for it is no other than the famous
Peak of Teneriffe, a volcano which is hardly burnt out yet,
and may burn up again any day, standing up out of the
sea more than twelve thousand feet high still, and once it
must have been double that height.   Some think that it

VOLCANO IN BARREN ISLAND, BAY OF BENGAL.

is perhaps the true Mount Atlas, which the old Greeks
named when first they ventured out of the Straits of
Gibraltar down the coast of Africa, and saw the great
peak far to the westward, with the clouds cutting off its

CRATER OF KILAUEA.

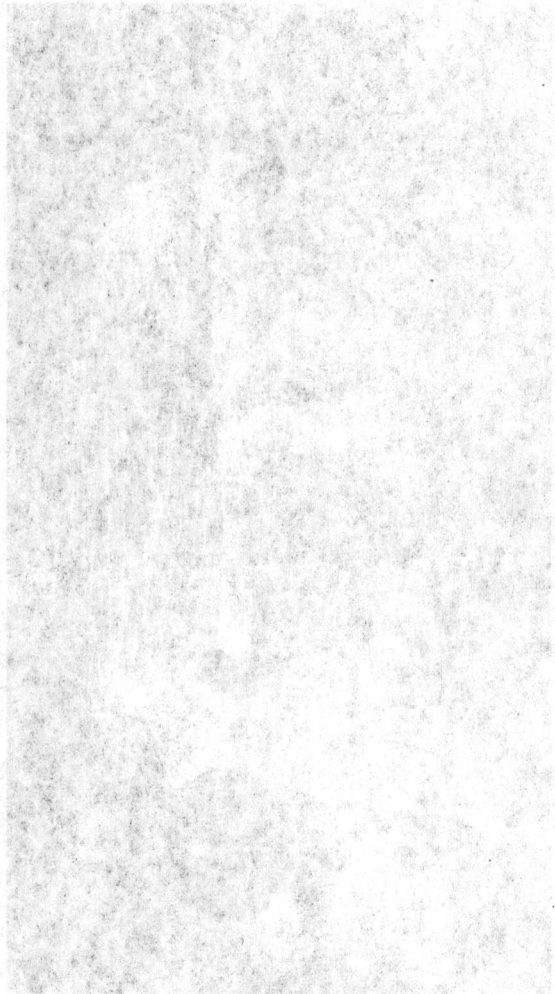

top; and said that it was a mighty giant, the brother
of the Evening Star, who held up the sky upon his
shoulders, in the midst of the Fortunate Islands, the
gardens of the daughter of the Evening Star, full of
strange golden fruits; and that Perseus had turned
him into stone, when he passed him with the Gorgon's
head.

But you will see, too, that most of these red and black
dots run in crooked lines; and that many of the clusters
run in lines likewise.

Look at one line: by far the largest on the earth.
You will learn a good deal of geography from it.

The red dots begin at a place called the Terribles, on
the east side of the Bay of Bengal. They run on, here
and there, along the islands of Sumatra and Java, and
through the Spice Islands; and at New Guinea the line
of red dots forks.

One branch runs south-east, through islands whose
names you never heard, to the Friendly Islands, and to
New Zealand. The other runs north, through the
Philippines, through Japan, through Kamtschatka; and
then there is a little break of sea, between Asia and
America; but beyond it, the red dots begin again in the
Aleutian Islands, and then turn down the whole west
coast of America, down from Mount Elias (in what was,
till lately, Russian America) towards British Columbia.
Then, after a long gap, there are one or two in Lower
California: and when we come down to Mexico we find
the red dots again plentiful, and only too plentiful; for

they mark the great volcanic line of Mexico, of which you
will read, I hope, some day, in Humboldt's works.

But the line does not stop there. After the little gap
of the Isthmus of Panama, it begins again in Quito, where
stand the huge volcanoes Chimborazo, Pasto, Antisanas,
Cotopaxi, Pichincha, Tunguragua—smooth cones from
fifteen to twenty thousand feet high, shining white with
snow, till the heat inside melts it off, and leaves the cinders
of which the peaks are made all black and ugly among the
clouds, ready to burst out in smoke and fire. South of
them, again, there is a long gap, and then another line
of red dots—Arequipa, Chipicani, Gualatieri, Atacama—
as high or higher than those in Quito. On the seashore
below those volcanoes stood the hapless city of Arica.
Then comes another gap; and then a line of more
volcanoes in Chili, at the foot of which happened that
fearful earthquake of 1835 (besides many more); and so
the line of dots runs down to the southernmost point of
America.

What a line we have traced! Long enough to go
round the world if it were straight. A line of holes
out of which steam, and heat, and cinders, and melted
stones are rushing up perpetually in one place and
another. Now the holes in this line which are near
each other have certainly something to do with each
other. For instance, when the earth shook in 1868 round
the volcanoes of Quito, it shook also round the volcanoes
of Peru, though they were six hundred miles away.
And there are many stories of earthquakes being felt, or

awful underground thunder heard, while volcanoes were breaking out hundreds of miles away. I will give you a very curious instance of that.

If you look at the West Indies on the map, you will see a line of red dots runs through the Windward Islands: there are two volcanoes in them, one in Guadaloupe, and one in St. Vincent (I will tell you a curious story, presently, about that last), and a little volcano, which now only sends out mud, in Trinidad. There the red dots stop: but then begins along the north coast of South America (which you must learn to call the Spanish Main) a line of mountain country called Cumana and Caraccas, which has often been horribly shaken by earthquakes.

Now once, when the volcano in St. Vincent began to pour out a vast stream of melted lava, a noise like thunder was heard underground, over thousands of square miles beyond those mountains, in the plains of Calabozo, and on the banks of the Apure, more than six hundred miles away from the volcano—a plain sign that there was something underground which joined them together, perhaps a long crack in the earth. Look for yourselves at the places, and you will see that (as Humboldt says) it is as strange as if an eruption of Mount Vesuvius was heard in the north of France.

So it seems as if these lines of volcanoes stood along cracks in the rind of the earth, through which the melted stuff inside was for ever trying to force its way; and that, as the crack got choked up in one place by the melted stuff cooling and hardening again into stone, it was burst in another place, and a fresh volcano made or an old one reopened.

Now we can understand why earthquakes should be most common round volcanoes ; and we can understand, too, why they would be worse before a volcano breaks out, because then the steam is trying to escape ; and we can understand, too, why people who live near volcanoes are glad to see them blazing and spouting, because then they have hope that the steam has found its way out, and will not make earthquakes any more for awhile. But still that is merely foolish speculation on chance. Volcanoes can never be trusted. No one knows when one will break out or what it will do ; and those who live close to them—as the city of Naples is close to Mount Vesuvius—must not be astonished if they are blown up or swallowed up, as that great and beautiful city of Naples may be without a warning, any day.

For what happened to that same Mount Vesuvius nearly eighteen hundred years ago, in the old Roman times ?

For ages and ages it had been lying quiet, like any other hill. Beautiful cities were built at its foot, filled with people who were as handsome, and as comfortable, and (I am afraid) as wicked, as people ever were on earth. Fair gardens, vineyards, oliveyards, covered the mountain slopes. It was held to be one of the Para-dises of the world. As for the mountain's being a burning mountain, who ever thought of that ? To be sure, on the top of it was a great round crater, or cup, a mile or more across, and a few hundred yards deep. But that was all overgrown with bushes and wild vines, full of boars and deer.

What sign of fire was there in that?

To be sure, also, there was an ugly place below by

BAY OF NAPLES.

the seashore, called the Phlegræan fields, where smoke and brimstone came out of the ground, and a lake called Avernus, over which poisonous gases hung, and which (old stories told) was one of the mouths of the Nether Pit. But what of that? It had never harmed any one, and how could it harm them?

F

So they all lived on, merrily and happily enough, till, in the year A.D. 79, there was stationed in the Bay of Naples a Roman Admiral, called Pliny, who was also a very studious and learned man, and author of a famous old book on natural history. He was staying on shore with his sister; and as he sat in his study she called him out to see a strange cloud which had been hanging for some time over the top of Mount Vesuvius. It was in shape just like a pine-tree; not, of course, like one of our branching Scotch firs here, but like an Italian stone pine, with a long straight stem and a flat parasol-shaped top. Sometimes it was blackish, sometimes spotted; and the good Admiral Pliny, who was always curious about natural science, ordered his cutter and went away across the bay to see what it could be.

Earthquake shocks had been very common for the last few days; but I do not suppose that Pliny had any notion that the earthquakes and the cloud had aught to do with each other. However, he soon found out that they had, and to his cost. When he got near the opposite shore some of the sailors met him and entreated him to turn back. Cinders and pumice-stone were falling down from the sky, and flames breaking out of the mountain above. But Pliny would go on. He said that if people were in danger it was his duty to help them, and that he must see this strange cloud, and note down the different shapes into which it changed.

The hot ashes fell faster and faster; the sea ebbed out suddenly, and left them nearly dry, and Pliny turned

DESTRUCTION OF POMPEII.

away to a place called Stabiæ, to the house of his friend
Pomponianus, who was just going to escape in a boat.
Brave Pliny told him not to be afraid, ordered his bath
like a true Roman gentleman, and then went in to dinner
with a cheerful face.   Flames came down from the mountain, nearer and nearer as the night drew on ; but Pliny
persuaded his friend that they were only fires in some
villages from which the peasants had fled, and then went
to bed and slept soundly.

However, in the middle of the night they found the
courtyard being fast filled with cinders, and, if they had
not woke up the Admiral in time, he would never have
been able to get out of the house.   The earthquake
shocks grew stranger and fiercer, till the house was ready
to fall ; and Pliny and his friend, and the sailors and the
slaves, all fled into the open fields, amid a shower of
stones and cinders, tying pillows over their heads to prevent their being beaten down.

By this time the day had come; but not the dawn,
for it was still pitch dark as night.   They went down to
their boats upon the shore, but the sea raged so horribly
that there was no getting on board them.   Then Pliny
grew tired, and made his men spread a sail for him, and
lay down on it.   But there came down upon them a rush
of flames and a horrible smell of sulphur, and all ran for
their lives.   Some of the slaves tried to help the Admiral
upon his legs ; but he sank down again overpowered with
the brimstone fumes, and so was left behind.   When they
came back again there he lay dead, but with his clothes

in order, and his face as quiet as if he had been only sleeping. And that was the end of a brave and learned man, a martyr to duty and to the love of science.

But what was going on in the meantime?

Under clouds of ashes, cinders, mud, lava, three of those happy cities were buried at once—Herculaneum, Pompeii, Stabiæ. They were buried just as the people had fled from them, leaving the furniture and the earthenware, even often jewels and gold, behind, and here and there among them a human being who had not had time to escape from the dreadful deluge of dust.

The ruins of Herculaneum and Pompeii have been dug into since; and the paintings, especially in Pompeii, are found upon the walls still fresh, preserved from the air by the ashes which have covered them in. In the famous museum of Naples, the curiosities which have been dug out of the ruined cities are preserved; and you may walk along the streets of Pompeii, and see the wheel tracks in the pavement along which carts and chariots rumbled two thousand years ago. If you go to the Crystal Palace, and to the Pompeian Court, as it is called, you will see an exact model of one of these old buried houses, copied even to the very paintings on the walls, and judge for yourself what sort of life these thoughtless, luckless people lived two thousand years ago.

And what had become of Vesuvius, the treacherous mountain?

Half or more than half of the side of the old crater had been blown away; and what was left, which is now called the Monte Somma, stands in a half circle round the new cone and new crater which is burning at this very day. After that eruption which killed Pliny, Vesuvius fell asleep again, and did not awake for a hundred and thirty-four years, and then again for two hundred and sixty-nine years; but it has been growing more and more restless as the ages have passed on, and now hardly a year passes without its sending out smoke and stones from its crater, and streams of lava from its sides.

VOLCANOES AND EARTHQUAKES.

CHAPTER VI.

THEIR CONSTRUCTION.

AND now, I suppose, you will want to know
What a volcano is like, and what a cone, and a
crater, and lava are ?

What a volcano is like is easy enough to show you ; for
it is the most simply and beautifully shaped of all mountains, and volcanoes are alike all over the world, whether
they be large or small. Almost every volcano in the world,
I believe, is, or has been once, of the shape which you
see in the drawing on page 75 ; even those volcanoes in
the Sandwich Islands, of which you have often heard,
which are now great lakes of boiling fire upon flat downs
without any cone to them at all.   They, I believe, are
volcanoes which have fallen in ages ago ; just as in Java
a whole burning mountain fell in in the night of the 11th

of August, in the year 1772. Then, after a short and terrible earthquake, a bright cloud suddenly covered the whole mountain. The people who dwelt around it tried to escape ; but before the poor souls could get away the earth sunk beneath their feet, and the whole mountain fell in, and was swallowed up, with a noise as if great cannon were being fired. Forty villages and nearly 8,000 people were destroyed, and where the mountain had been was only a plain of red-hot stones. In the same way, in the year 1698, the top of a mountain in Quito fell in in a single night, leaving only two immense peaks of rock behind, and pouring out great floods of mud mixed with dead fish; for there are underground lakes among those volcanoes, which swarm with little fish which never see the light.

But most volcanoes, as I say, are, or have been, the shape of the one which you see on page 75. This is the cone of Cotopaxi, in Quito, more than nineteen thousand feet in height. All those sloping sides are made of cinders and ashes, braced together, I suppose, by bars of solid lava stone inside, which prevent the whole from crumbling down. The upper part of the picture, you see, is white with snow as far down as a line which is fifteen thousand feet above the sea. For the mountain is in the tropics, close to the equator, and the snow will not lie in that hot climate any lower down. But now and then the snow melts off, and rushes down the mountain side in floods of water and of mud, and the cindery cone of Cotopaxi stands out black and dreadful against the clear blue sky, and then the people of that country know what is coming. The mountain is growing so hot inside that it melts off its snowy covering; and soon it will burst forth with smoke and steam and red-hot stones, and earthquakes which will shake the ground, and roars that will be heard it may be hundreds of miles away.

And now for the words, cone, crater, lava. If I can make you understand those words, you will see why volcanoes must be in general of the shape of Cotopaxi.

Cone, crater, lava: those words make up the alphabet of volcano learning.

VOLCANO, OLD AND NEW CRATER.

The cone is the outside of a huge chimney.

The crater is the mouth of it.

The lava is the ore which is being melted in the furnace below, that it may flow out over the surface of the old land, and make new land instead.

And where is the furnace itself?

Who can tell that? Under the roots of the mountains, under the depths of the sea; down "the path which no fowl knoweth, and which the vulture's eye hath not seen, the lion's whelp hath not trodden it, nor the fierce lion

COTOPAXI.

passed by it. There He putteth forth His hand upon the rock ; He overturneth the mountains by the roots ; He cutteth out rivers among the rocks ; and His eye seeth every precious thing "—while we, like little ants, run up and down outside the earth, scratching, like ants, a few feet down, and calling that a deep ravine ; or peeping a few feet down into the crater of a volcano, unable to guess what precious things may lie below—below even the fire which blazes and roars up through the thin crust of the earth. For of the inside of this earth we know nothing whatsoever. We only know that it is, on an average, several times as heavy as solid rock ; but how that can be we know not.

So let us look at the chimney, and what comes out of it, for we can see very little more.

Why is a volcano like a cone ?

For the same cause for which a molehill is like a cone, though a very rough one ; and that the little heaps which the burrowing beetles make on the moor, or which the ant-lions in France make in the sand, are all something of the shape of a cone, with a hole like a crater in the middle. What the beetle and the ant-lion do on a very little scale, the steam inside the earth does on a great scale. When once it has forced a vent into the outside air, it tears out the rocks underground, grinds them small against each other, often into the finest dust, and blasts them out of the hole which it has made. Some of them fall back into the hole, and are shot out again ; but most of them fall round the hole, most of them close to it, and fewer of them further off, till they are piled up in a ring round it, just as the sand is piled up round a beetle's burrow.

For days and weeks and months this goes on, even it may be for hundreds of years, till a great cone is

CRATER OF EXTINCT VOLCANO, INTERIOR.

formed round the steam vent, hundreds or thousands of feet in height, of dust and stones, and of cinders likewise. For recollect, that when the steam has blown away the cold earth and rock near the surface of the ground, it begins blowing out the hot rocks down below, red-hot, white-hot, and at last actually melted. But these, as they are hurled into the cool air above, become ashes, cinders, and blocks of stone again, making the hill on which they fall bigger and bigger continually. And thus volcanoes stand in no need of bricklayers, but make their chimneys build themselves.

CRATER OF EXTINCT VOLCANO, EXTERIOR.

And why is the mouth of the chimney called a crater?

Crater, as you know, is Greek for a cup. And the mouths of these chimneys, when they have become choked and stopped working, are often just the shape of a cup, or (as the Germans call them) kessels, which means kettles, or caldrons. I have seen some of them as

beautifully and exactly rounded as if a cunning engineer
had planned them, and had them dug out with the spade.
At first, of course, their sides and bottom are nothing but
loose stones, cinders, slag, ashes, such as would be thrown
out of a furnace.

But Nature, whenever she makes an ugly desolate place,
always tries to cover over its ugliness, and set something
green to grow over it, and make it pretty once more. I
have seen worn-out craters covered with short sweet turf
like so many chalk downs. I have seen them, too, filled
with bushes, which held woodcocks and wild boars.
Once I came on a beautiful round crater on the top of a
mountain, which at the bottom furnished soil to grow a
splendid crop of potatoes. And often worn-out craters
are turned into beautiful lakes. There are many such
crater-lakes in Italy, as you will see if ever you go there.
You recollect Lord Macaulay's ballad, "The Battle of the
Lake Regillus?" Then that Lake Regillus (if I recollect
right) is one of these round crater lakes. Many such
deep clear blue lakes have I seen in the Eifel, in Germany, and many a curious plant have I picked on their
shores, where once the steam blasted and the earthquake
roared, and the ash-clouds rushed up high into the
heaven, and buried all the land around in dust, which
is now fertile soil.

And long did I puzzle to find out why the water stood
in some craters, while others within a mile of them,
perhaps, were perfectly dry. That I never found out for
myself. But learned men tell me that the ashes which

fall back into the crater, if the bottom of it be wet from rain, will sometimes "set" (as it is called) into a hard cement; and so make the bottom of the great bowl water-proof, as if it were made of earthenware.

But what gives the craters this cup-shape at first?

Think. While the steam and stones are being blown out, the crater is an open funnel, with more or less upright walls inside. As the steam grows weaker, fewer and fewer stones fall outside, and more and more fall back again inside. At last they quite choke up the bottom of the great round hole. Perhaps, too, the lava or melted rock underneath cools and grows hard, and that chokes up the hole lower down. Then, down from the round edge of the crater the stones and cinders roll inward more and more. The rains wash them down, the wind blows them down. They roll to the middle, and meet each other, and stop. And so gradually the steep funnel becomes a round cup.

You may prove for yourself that it must be so, if you will try. Do you not know that if you dig a round hole in the ground, and leave it to crumble in, it is sure to become cup-shaped at last, though at first its sides may have been quite upright, like those of a bucket? If you do not know, get a trowel and make your little experiment.

And now you ought to understand what "cone" and "crater" mean. And more, if you will think for your-self, you may guess what would come out of a volcano when it broke out "in an eruption," as it is usually called.

G

First, clouds of steam and dust (what you would call smoke).

Then volleys of stones, some cool, some burning hot; and at the last, because it lies lowest of all, the melted rock itself, which is called lava.

And where would that come out? At the top of the chimney? At the top of the cone?

No. As I told you, the chimney of the furnace makes itself; the furnace door also makes itself.

The melted lava rises in the crater—the funnel inside the cone—but it never gets to the top. It is so enormously heavy that the sides of the cone cannot bear its weight, and give way low down. And then, through ashes and cinders, the melted lava burrows out, twisting and twirling like an enormous fiery earthworm, till it gets to the air outside, and runs off down the mountain in a stream of fire. And so you may see two eruptions at once—one of burning stones above, and one of melted lava below.

Now look at this figure. It represents a section of a volcano; that is, one cut in half to show you the inside. A is the cone of cinders. B, the black line up through the middle, is the funnel, or crack, through which steam, ashes, lava, and everything else rises. C is the crater mouth. D D D, which looks broken, are the old rocks which the steam heaved up and burst before it could get out. And what are the black lines across, marked E E E? They are the streams of lava which have burrowed out, some covered up again in cinders, some lying bare in the open air, some still inside the cone bracing it together, holding it up. Something like this is the inside of a volcano.

And what is lava?

That, I think, I must tell you another time, when I speak of the uses of volcanoes. But if you want to know (as I dare say you do) what the eruption of lava is like, you may read what follows concerning an eruption

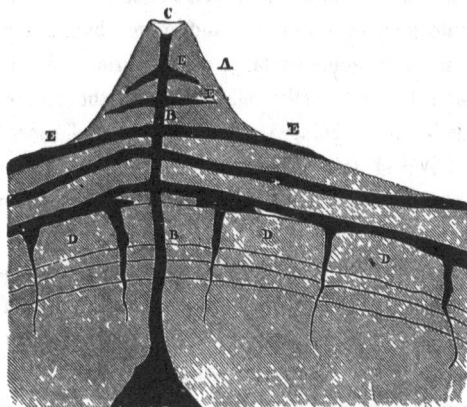

SECTION OF VOLCANO.

on Vesuvius, not from the main crater, but from a small one which had risen up suddenly on the outside of it.

This new cone, says the observer, was about two hundred feet high, and perhaps eighty or a hundred feet across at the top. And as he stood below it (it was not safe to go up it) smoke rolled out from its top, "rosy pink below," from the glare of the caldron, and above "faint

greenish or bluish silver of indescribable beauty, from the
light of the moon." But more. By good chance, the
cone began to send out, not smoke only, but brilliant
burning stones. "Each explosion," he says, "was like
a vast girandole of rockets, with a noise (such as rockets
would make) like the waves on a beach, or the wind
blowing through shrouds.

"The mountain was trembling the whole time. So it
went on for two hours and more; sometimes eight or
ten explosions in a minute, and more than a thousand
stones in each, some as large as two bricks end to end.
The largest ones mostly fell back into the crater; but
the smaller ones, being thrown higher, and more acted
on by the wind, fell in immense numbers on the leeward
slope of the cone" (of course, making it bigger and
bigger, as explained already), and of course, as they
were intensely hot and bright, making the cone look as
if it too was red-hot. But it was not so, he says, really.
The colour of the stones was rather "golden, and they
spotted the black cone over with their golden showers,
the smaller ones stopping still, the bigger ones rolling
down and jumping along just like hares."

"A wonderful pedestal," he says, "for the explosion
which surmounted it." How high the stones flew up
he could not tell. "There was generally one which
went much higher than the rest, and pierced upwards
towards the moon, who looked calmly down, mocking
such vain attempts to reach her." The large stones, of
course, did not rise so high; and some, he says, "only

just appeared over the rim of the cone, above which they came floating leisurely up, to show their brilliant forms and intense white light for an instant, and then subside again."

Try and picture that to yourselves, remembering that this was only a little side eruption, of no more importance to the whole mountain than the fall of a slate off the roof is of importance to the whole house. And then think how mean and weak man's fireworks, and even man's heaviest artillery, are, compared with the terrible beauty and terrible strength of the artillery underneath our feet.

VOLCANOES AND EARTHQUAKES

Part II.

WHY THEY ARE.

VOLCANOES AND EARTHQUAKES.

CHAPTER I.

LAND-MAKERS.

HAVE the tidings of the destruction, the often terrible destruction, of earthquakes and volcanoes ever puzzled you and made you sad? Do you ask in painful wonder why God permits such events?

Alas! who am I that I should answer you that?

But have you done wrong in asking me? No. You have asked me because you are a human being and a child of God, and not merely a cleverer sort of animal, an ape who can read and write and cast accounts. Therefore it is that you cannot be content, and ought not to be content, with asking how things happen, but must go on to ask why. You cannot be content, however, with knowing the causes

of things. Even if you knew all the natural science that
ever was or ever will be known to men, that would not
satisfy you, for it would only tell you the *causes* of
things, while your souls want to know the *reason* of
things besides ; and though I cannot tell you the reasons
of things, yet I believe that somehow, somewhen, somewhere, you will learn something of the *reason* of things.
For that thirst to know *why* was put into the heart by God
himself, and I believe that God would never have given
that thirst, if He had not meant to satisfy it.

But has it not occurred to some to ask themselves
whether it is not just possible that these ugly affairs are
sent among us for some beneficent purpose, or at all
events that they may form part and parcel of some great
scheme of providential arrangement which is at work for
good, and not for ill.

A ship sometimes strikes on a rock, and all on board
perish ; a railway train runs into another, or breaks down,
and then wounds and contusions are the order of the day ;
but nobody doubts that navigation and railway communication are great blessings.

None of the great natural provisions for producing good
are exempt in their workings from producing occasional
mischief. Storms disperse and dilute pestilential vapours,
and lightnings decompose and destroy them ; but both the
one and the other often annihilate the works of man, and
inflict upon him sudden death.

Well, then, I think I shall be able to show that the
volcano and the earthquake, dreadful as they are, as local

SEA ON COAST AT ST. MALO.

and temporary visitations, are in fact unavoidable (I had almost said necessary) incidents in a vast system of action to which we owe the very ground we stand upon, the very land we inhabit, without which neither man. beast, nor bird would have a place for their existence, and the world would be the habitation of nothing but fishes.

Now, to make this clear, I must go a little out of my way and say something about the first principles of geology.

Geology does not pretend to go back to the creation of the world, or concern itself about its primitive state, but it does concern itself with the changes it sees going on in it now, and with the evidence of a long series of such changes it can produce in the most unmistakable features of the structure of our rocks and soil, and the way in which they lie one on the other.

As to what we SEE *going on.*—We see everywhere, and along every coast-line, the sea warring against the land, and everywhere overcoming it, wearing and eating it down, and battering it to pieces, grinding those pieces to powder, carrying that powder away, and spreading it out over its own bottom, by the continued effect of the tides and currents.

Look at our chalk cliffs, which once, no doubt, extended across the Channel to the similar cliffs on the French coast.

What do we see ?

Precipices cut down to the sea-beach, constantly hammered by the waves and constantly crumbling, the beach

itself made of the flints outstanding after the softer chalk
has been ground down and washed away, themselves
grinding one another under the same ceaseless discipline,
first rounded into pebbles, then worn into sand, and then
carried out farther and farther down the slope, to be
replaced by fresh ones from the same source.

Upon every mountain-top, and on every level plain,

CHALK CLIFFS.

rain is carrying on the same destructive work; on hard
rock slowly, on softer material swiftly, carrying land into
rivers, and by rivers, into the sea, sometimes to hundreds
of miles away from the river's mouth.

Well, the same thing is going on *everywhere, round every
coast* of Europe, Asia, Africa, and America. Foot by foot
or inch by inch, month by month or century by century,
down everything MUST *go.* Time is as nothing in geology.

And what the sea is doing the rivers are helping it to do. Look at the sandbanks at the mouth of the Thames. What are they but the materials of our island carried out to sea by the stream? The Ganges carries away from the

RAIN WASHINGS, MOUNTAIN PEAK.

soil of India, and delivers into the sea, as much solid substance *daily* as is contained in the great pyramid of Egypt. The Irawaddy sweeps off from Burmah sixty-two cubic feet of earth in every second of time on an average, and there are eighty-six thousand four hundred seconds in every day, and three hundred and sixty-five days in every year, and so on for the other rivers.   What has

RIVER CUTTINGS IN NORWAY.

become of all that great bed of chalk which once covered all the weald of Kent, and formed a continuous mass from Ramsgate and Dover to Beechy Head, running inland to Madamscourt Hill and Sevenoaks.   All clean gone, and

swept out into the bosom of the Atlantic, and there forming other chalk-beds.

Now, geology assures us, on the most conclusive and

RIVER CUTTINGS IN COLORADO.

undeniable evidence, that ALL our present land, all our
continents and islands, have been formed in this way out
of the ruins of former ones. The old ones which existed
at the beginning of things have all perished, and what we

H

now stand upon has most assuredly been, at one time or other, perhaps many times, the bottom of the sea.

Well, then, there is power enough at work, and it has been at work long enough, utterly to have cleared away and spread over the bed of the sea all our present existing continents and islands, had they been placed where they are at the creation of the world; and from this it follows, as clear as demonstration can make it, that without *some* process of renovation or restoration to act in antagonism to this destructive work of old Neptune, there would not now be remaining a foot of dry land for living thing to stand upon.

Now, what *is* this process of restoration?

Let the volcano and the earthquake tell their tale. Let the earthquake tell how, within the memory of man— under the eyesight of eye-witnesses, one of whom (Mrs. Graham) has described the fact—the whole coast-line of Chili, for a hundred miles about Valparaiso, with the mighty chain of the Andes—mountains to which the Alps shrink into insignificance—was hoisted at one blow (in a single night, Nov. 19, A.D. 1822) from two to seven feet above its former level, leaving the beach below the old low water-mark high and dry, leaving the shellfish sticking on the rocks out of reach of water, leaving the seaweed rotting in the air, or rather drying up to dust under the burning sun of a coast where rain never falls.

The ancients had a fable of Titan hurled from heaven and buried under Ætna, and by his struggles causing the earthquakes that desolated Sicily. But here we have an exhibition of Titanic forces on a far mightier scale. One of the Andes uplifted on this occasion was the gigantic mass of Aconcagua, which overlooks Valparaiso.

To bring home to the mind the conception of such an effort, we must form a clear idea of what sort of mountain this is. It is nearly 24,000 feet in height. Chimborazo, the loftiest of the volcanic cones of the Andes, is lower by 2,500 feet; and yet Ætna, with Vesuvius at the top of it, and another Vesuvius piled on that, *would little more than surpass the midway height of the snow-covered portion of Chimborazo*. On the occasion I am speaking of, at least 10,000 square miles of country were estimated as having been upheaved, and the upheaval was not confined to the land, but extended far away to sea, which was proved by the soundings off Valparaiso and along the coast having been found considerably shallower than they were before the shock.

Again, in the year 1819, in an earthquake in India, in the district of Cutch, bordering on the Indus, a tract of country more than fifty miles long and sixteen broad was suddenly raised ten feet above its former level. The raised portion still stands up above the unraised, like a long perpendicular wall, which is known by the name of the " Ullah Bund," or " God's Wall."

Again, in 1538, in that convulsion which threw up the Monte Nuovo (New Mountain), a cone of ashes four hundred and fifty feet high, in a single night, the whole coast of Pozzuoli, near Naples, was raised twenty feet above its former level, and remains so permanently upheaved to this day. And I could mention innumerable other instances of the same kind.*

This, then, is the manner in which the earthquake does its work; *and it is always at work.* Somewhere or other

SEA-BORINGS.

in the world, there is, perhaps, not a day, certainly not a month, without an earthquake.

* Not that earthquakes always *raise* the soil; there are plenty of instances of subsidence. &c.

In those districts of South and Central America, where the great chain of volcanic cones is situated—Chimborazo, Cotopaxi, and a long list with names unmentionable, or at least unpronounceable—the inhabitants no more think of

WATER-SAW.

counting earthquake shocks than we do of counting showers of rain. Indeed, in some places along that coast, a shower is a greater rarity. Even in our own island, near Perth, a year seldom passes without a shock, happily, within the records of history, never powerful enough to do any mischief.

It is not everywhere that this process goes on by fits and starts.

For instance, the northern gulfs, and borders of the Baltic Sea, are steadily shallowing, and the whole mass of Scandinavia, including Norway, Sweden, and Lapland, is rising out of the sea at the average rate of about two feet per century. But as this fact (which is perfectly well established by reference to ancient high and low water-marks) is not so evidently connected with the action of earthquakes, I shall not further refer to it just now. All that I want to show is, that there is a great cycle of changes going on, in which the earthquake and volcano act a very conspicuous part, *and that part a restorative and conservative one*, in opposition to the steadily destructive and levelling action of the ocean waters.

How this can happen, what can be the origin of such an enormous power thus occasionally exerting itself, will no doubt seem very marvellous—little short, indeed, of miraculous intervention—but the mystery, after all, is not quite so great as at first it seems.

We are permitted to look a little way into these great secrets of nature ; not far enough, indeed, to clear up every difficulty, but quite enough to penetrate us with admiration of that wonderful system of counterbalances and compensations, that adjustment of causes and conse-quences, by which, throughout all nature, evils are made to work their own cure. life to spring out of death, and

renovation to tread in the steps and efface the vestiges
of decay.

The key to the whole affair is to be found in the central
heat of the earth. This is a fact established by direct
evidence up to a certain point, and standing out from
plain facts as a matter of unavoidable conclusion in a
hundred ways.

We all know that when we go into a cellar out of a
summer sun it feels *cool*, but when we go into it out of a
wintry frost it is *warm*. The fact is, that a cellar, or a
well, or any pit of a moderate depth, has always, day and
night, summer and winter, the same degree of warmth,
the same *temperature*, as it is called, and *that* always and
everywhere is the same, or nearly the same, as the
average warmth of the climate of the place.

Forty or fifty feet deep in the ground, a thermometer
here, in this spot,* would always mark the same degree,
49° that is, or seventeen degrees above the freezing point.
Under the equator, at the same depth, it always stands
at 84°, which is our *hot summer* heat, but which there is
the average heat of the whole year. And this is so
everywhere.

Just at the surface, or a few inches below it, the
ground is warm in the daytime, cool at night ; at two or
three feet deep the difference of day and night is hardly
perceptible, but that of summer and winter is considerable. But at forty or fifty feet this difference also disappears, and you find a perfectly fixed, uniform degree

* At Hawkhurst, in Kent.

of warmth, day and night, summer and winter, year after
year.

But when we go deeper, as, for instance, down into
mines or coal-pits, this one broad and general fact is

PEEP INTO THE INTERIOR.

always observed—everywhere, in all countries, in all
latitudes, in all climates, wherever there are mines or
deep subterranean caves, the deeper you go, the hotter
the earth is found to be.

In one and the same mine, each particular depth has its own particular degree of heat, which never varies, but the lower always the hotter; and that not by a trifling, but what may well be called an astonishingly rapid rate of

GEYSER.

increase—about a degree of the thermometer additional warmth for every ninety feet of additional depth, which is about 58° per mile!—so that, if we had a shaft sunk a mile deep, we should find a heat of 105°, which is much hotter than the hottest summer day ever experienced in England.

It is not everywhere, however, that it is worth while to sink a shaft to any great depth; but borings for water (in what are called Artesian wells) are often made to enormous depths, and the water always comes up hot; and the deeper the boring, the hotter the water.

There is a very famous boring of this sort in Paris, at La Grenelle. The water rises from a depth of 1,794 feet, and its temperature is 82° of our scale, which is almost that of the equator. Again, at Salzwerth, in Oeynhausen, in Germany, in a boring for salt-springs 2,144 feet deep, the salt water comes up with a still higher heat, viz. 91°.

Then, again, we have natural hot-water springs, which rise, it is true, from depths we have no means of ascertaining, but which, from the earliest recorded times, have always maintained the same heat. At Bath, for instance, the hottest well is 117° Fahr. On the Arkansas River, in the United States, is a spring of 180°, which is scalding hot; and that out of the neighbourhood of any volcano.

Now, only consider what sort of a conclusion this lands us in.

VOLCANOES AND EARTHQUAKES.

CHAPTER II.

LAND-MAKERS.—(*Continued.*)

THIS globe of ours is eight thousand miles in diameter; a mile deep on its surface is a mere scratch. If a man had twenty greatcoats on, and I found under the first a warmth of 60° above the external air, I should expect to find 60° more under the second, and 60° more under the third, and so on; and, within all, *no man*, but a mass of red-hot iron. Just so with the outside crust of the earth.

Every mile thick is such a great coat, and at twenty miles' depth, according to this rate, the ground must be fully red-hot; and at no such very great depth beyond, either the whole must be melted, or only the most infusible and intractable kinds of material, such as our fire-clays and flints, would present some degree of solidity.

Well, now, the land is perpetually wearing down, and

the materials carried out to sea.   The coat of heavier
matter is thinning off towards the land, and thickening
over all the bed of the sea.

What must happen ?

If a ship floats even on her keel, transfers weight from
the starboard to the larboard side, will she continue to
float even ?   No, certainly.   She will heel over to larboard.   Many a good ship has gone to the bottom in this
way.   If the continents are lightened, they will rise ; if
the bed of the sea receives additional weight, it will sink.
The bottom of the Pacific *is* sinking, in point of fact.

Not that the Pacific is becoming *deeper*.   This seems a
paradox ; but it is easily explained.   The whole bed of
the sea is in the act of being pressed down *by the laying
on of new solid substance over its bottom*.   The new bot'om
then is laid upon the old, and so the actual bed of the
ocean remains at the same distance from the surface
water.

But what becomes of the Pacific islands ?   They form
part and parcel of the old bottom ; and Dr. Darwin has
shown, by the most curious and convincing proofs, that
they *are sinking*, and *have been sinking for ages*, and are
only kept above water—by what, think you ?   By the
labours of the coral insects, which always build up to the
surface !

It is impossible but that this increase of pressure in
some places and relief in others must be very unequal in
their bearings.   So that at some places or other this solid
floating crust must be brought into a state of strain, and

if there be a weak or a soft part, a crack will at last take place. When this happens, down goes the land on the heavy side, and up on the light side.

Now this is exactly what took place in the earthquake which raised the "Ullah Bund" in Cutch. I have told you of a great crack down across the country, not far from the coast-line; the inland country rose ten feet, but much of the sea-coast, and probably a large tract in the bed of the Indian Ocean, sunk considerably below its former level. And just as you see when a crack takes place in ice the water oozes up, so this kind of thing is always, or almost always, followed by an upburst of the subterranean fiery matter. The earthquake of Cutch was terminated by the outbreak of a volcano at the town of Bhooi, which it destroyed.

Now where, following out this idea, should we naturally expect such cracks and outbreaks to happen?

Why, of course, along those lines where the relief of pressure on the land side is the greatest, and also its increase on the sea side; that is to say, along or in the neighbourhood of the sea-coasts, where the destruction of the land is going on with most activity.

Well, now, it is a remarkable fact in the history of volcanoes, that there is hardly an instance of an active volcano at any considerable distance from the sea-coast. All the great volcanic chain of the Andes is close to the western coast-line of America. Ætna is close to the sea; so is Vesuvius; Teneriffe is very near the African coast; Mount Erebus is on the edge of the great Antarctic continent. Out of two hundred and twenty-five volcanoes which are known to have been in actual eruption over the whole earth within the last hundred and fifty years, I remember only a single instance of one more than three hundred and twenty miles from the sea, and even *that* is on the edge of the Caspian, the largest of all the inland seas—I mean Mount Demawend, in Persia.

Suppose from this, or from any other cause, a crack to take place in the solid crust of the earth. Don't imagine that the melted matter below will simply ooze up quietly, as water does from under an ice-crack. No such thing. There is an element in the case we have not considered : steam and condensed gases.

We all know what happens when a crack takes place in a high-pressure steam-boiler, with what violence the contents escape, and what havoc takes place. Now there is no doubt that among the minerals of the subter-ranean world, there is water in abundance, and sulphur and many other vaporisable substances all kept subdued and repressed by the enormous pressure. Let this pres-sure be relieved, and forth they rush, and the nearer they approach the surface the more they expand, and the greater is the explosive force they acquire, till at length, after more or fewer preparatory shocks, each accompanied with progressive weakening of the overlying strata, the surface finally breaks up, and forth rushes the imprisoned power, with all the awful violence of a volcanic eruption.

Certainly a volcano does seem to be a very bad neigh-

MOUNT EREBUS.

bour; and yet it does afford a compensation in the extraordinary richness of the volcanic soil, and the fertilising quality of the ashes thrown out.

The flanks of Somma (the exterior crater of Vesuvius) are covered with vineyards producing wonderful wine, and whoever has visited Naples will not fail to be

VESUVIUS AT NIGHT.

astonished at the productiveness of the volcanised territory as contrasted with the barrenness of the limestone rocks bordering on it. There you will see the amazing sight (as an English farmer would call it) of a triple crop growing at once on the same soil—a vineyard,

I

an orchard, and a cornfield all in one. A magnificent wheat crop, five or six feet high, overhung with clustering grape-vines swinging from one apple or pear tree to another in the most luxuriant festoons !

An earthquake travels from place to place; how does it travel ?

It runs along the earth precisely in the same manner, and according to the same mechanical laws, as a wave along the sea, or rather as the waves of sound run along the air, but quicker.

The earthquake which destroyed Lisbon ran out from thence, as from a centre, in all directions, at a rate averaging about twenty miles per minute, as far as could be gathered from a comparison of the times of its occurrence at different places; but there is little doubt that it must have been retarded by having to traverse all sorts of ground, for a blow or shock of any description is conveyed through the substance on which it is delivered with the rapidity of *sound in* that substance.

Perhaps it may be new to many to be told that sound is conveyed by water, by stone, by iron, and indeed by everything, and at a different rate for each. In air it travels at the rate of about 1,140 feet per second, or about twelve miles in a minute. In water much faster, more than four times as fast (4,700 feet). In iron ten times as fast (11,400 feet), or about a hundred and thirty miles in a minute; so that a blow delivered (endways) at one end of an iron rod, a hundred and thirty miles long, would only reach the other after the lapse of a minute, and a pull at one end of an iron wire of that length would require a minute before it would be felt at the other.

But the substance of the earth through which the shock is conveyed is not only far less elastic than iron, but it does not form a coherent, connected body ; it is full of interruptions, cracks, loose materials, and all these tend to deaden and retard the shock ; and putting together all the accounts of all the earthquakes that have been exactly observed, their rate of travel may be taken to vary from as low as twelve or thirteen miles a minute to seventy or eighty, but perhaps the low velocities arise from oblique waves.

The way, then, that we may conceive an earthquake to travel is this. I shall take the case which is most common when the motion of the ground to and fro is horizontal. *How far* each particular spot on the surface of the ground is actually pushed from its place there is no way of ascertaining, since all the surrounding objects receive the same impulse almost at the same instant of time, but there are many indications that it is often several yards.

In the earthquake of Cutch, which I have mentioned, trees were seen to flog the ground with their branches, which proves that their stems must have been jerked suddenly away for some considerable distance and as suddenly pushed back ; and the same conclusion follows from the sudden rise of the water of lakes on the side where the shock reaches them, and its fall on the opposite side ; the bed of the lake has been jerked away for a certain distance from under the water and pulled back.

Now, suppose a row of sixty persons, standing a mile apart from each other, in a straight line, in the direction in which the shock travels, at a rate, we will suppose, of sixty miles per minute, and let the ground below the first get a sudden and violent shove, carrying it a yard in the direction of the next. Since this shock will not reach the next till after the lapse of one second of time, it is clear that the space between the two will be shortened by a yard, and the ground—that is to say, not the mere loose soil on the surface, but the whole mass of solid rock below down to an unknown depth—compressed, or driven into a smaller space. It is this compression that carries the shock forwards. The elastic force of the rocky matter, like a coiled spring, acts both ways; it drives back the first man to his old place, and shoves the second a yard nearer to the third, and so on.

Instead of men, place a row of tall buildings or columns, and they will tumble down in succession, the base flying forwards, and leaving the tops behind to drop on the soil on the side *from* which the shock came.

This is just what was seen to happen in Messina in the great Calabrian earthquake. As the shock ran along the ground, the houses of the Faro were seen to topple down in succession, beginning at one end and running on to the other, as if a succession of mines had been sprung. In the earthquake in Cutch, a sentinel, standing at one end of a long straight line of wall, saw the wall bow forward and recover itself, not all at once, but with a swell like a wave running all along it with immense rapidity.

In this case it is evident that the earthquake wave must have had its front oblique to the direction of the wall (just as an obliquely-held ruler runs along the edge of a page of paper while it advances, like a wave of the sea, perpendicularly to its own length).

In the study of these vast and awful phenomena we are brought in contact with those immense and rude powers of nature which seem to convey to the imagination the impress of brute force and lawless violence; but it is not so.

Such an idea is not more derogatory to the wisdom and benevolence that prevails throughout all the scheme of creation than it is in itself erroneous. In their wildest paroxysms the rage of the volcano and the earthquake is subject to great and immutable laws: they feel the bridle and obey it. The volcano bellows forth its pent-up over-plus of energy, and sinks into long and tranquil repose. The earthquake rolls away, and industry, that balm which nature knows how to shed over every wound, effaces its traces, and festoons its ruins with flowers. There is mighty and rough work to be accomplished, and it cannot be done by gentle means.

It seems, no doubt, terrible, awful, perhaps harsh, that twenty or thirty thousand lives should be swept away in a moment by a sudden and unforeseen calamity; but we must remember that sooner or later every one of those lives must be called for, and it is by no means the most sudden end that is the most afflictive.

It is well, too, that we should contemplate occasionally,

if it were only to teach us humility and submission, the immense energies which are everywhere at work in maintaining the system of nature we see going on so smoothly and tranquilly around us, and of which these furious outbreaks, after all, are but minute and for the moment unbalanced surpluses in the great account.

The energy requisite to overthrow a mountain is as a drop in the ocean compared with that which holds it in its place, and makes it a mountain.

Chemistry tells us that the forces constantly in action to maintain four grains of zinc in its habitual state, when only partially and sparingly let loose in the form of electricity, would supply the lightning of a considerable thunderstorm.

And we learn from optical science that in even the smallest element of every material body, nay, even in *what we call* empty space, there are forces in perpetual action to which even such energies sink into insignificance.

Yet, amid all this, nature holds her even course : the flowers blossom ; animals enjoy their brief span of existence ; and man has leisure and opportunity to contemplate and adore, secure of the watchful care which provides for his well-being at every instant that he is permitted to remain on earth.

VOLCANOES AND EARTHQUAKES.

CHAPTER III.

MAN-MAKERS.

WHY, you asked, are there such terrible things as volcanoes? Of what use can they be?

We have seen their use as restorers of lost land, repairing the wastes of the billows of time and the deep.

They are of many more uses, more, doubt not, than we know as yet, or ever shall know. But of one of their uses I can tell you. They make, or help to make, divers and sundry curious things, from gunpowder to your body and mine.

What! our bodies?

You can understand their helping to make gunpowder, because the sulphur in it is often found round volcanoes; and you recollect the story of the brave Spaniard, who, when his fellows wanted materials for gunpowder, had himself lowered in a basket down the crater of a South American volcano, and gathered sulphur for them off the burning cliffs.

But how can volcanoes help to make you? Are you made of lava? Or is there lava in you?

I did not say that volcanoes helped to make *you*. I said that they helped to make your body; which is a very different matter, as I beg you to remember, now and always. Your body is no more you yourself than the hoop which you trundle or the pony which you ride. It is, like them, your servant, your tool, your instrument, your organ, with which you work; and a very useful, trusty, cunningly-contrived organ it is, and therefore I advise you to make good use of it, for you are responsible for it.

But you yourself are not your body, or your brain, but something else, which we call your soul, your spirit, your life. And that " you yourself" would remain just the same if it were taken out of your body, and put into the body of a bee, or of a lion, or any other body, or into no body at all. At least so I believe; and so, I am happy to say, nine hundred and ninety-nine thousand nine hundred and ninety-nine people out of every million have always believed, because they have used their human instincts and their common sense, and have obeyed (without knowing it) the warning of a great and good philosopher called Herder, that " the organ is in no case the power which works by it;" which is as much as to say, that the engine is not the engine-driver, nor the spade the gardener.

There have always been, and always will be, a few people who cannot see that. They think that a man's soul is part of his body, and that he himself is not one thing, but a great number of things. They think that his mind and character are only made up of all the thoughts, and feelings, and recollections which have passed through his brain; and that as his brain changes, he himself must change, and become another person again, continually. But do not you agree with them, but keep in mind wise Herder's warning that you are not to "confound the organ with the power," or the engine with the driver, or your body with yourself; and then we will go on and consider how a volcano, and the lava which flows from it, helps to make your body.

Now I know that the Scotch have a saying, "That you cannot make broth out of whinstones" (which is their name for lava). But, though they are very clever people, they are wrong there. I never saw any broth in Scotland, as far as I know, but what whinstones had gone to the making of it; nor a Scotch boy who had not eaten many a bit of whinstone, and been all the better for it.

Of course, if you simply put the whinstones into a kettle and boiled them, you would not get much out of them by such rough cookery as that. But Nature is the best and most delicate of all cooks; and she knows how to pound, and soak, and stew whinstones so delicately, that she can make them sauce and seasoning for meat, vege-tables, puddings, and almost everything that you eat;

and can put into your veins things which were spouted up red-hot by volcanoes, ages and ages since, perhaps at the bottom of ancient seas which are now firm dry land.

This is very strange, and you would think it stranger still if you had ever seen the flowing of a lava stream.

Out of a cave of slag and cinders in the black hillside rushes a golden river, flowing like honey, and yet so tough that you cannot thrust a stick into it, and so heavy that great stones (if you throw them on it) float on the top, and are carried down like corks on water. It is so hot that you cannot stand near it more than a few seconds—hotter, perhaps, than any fire you ever saw; but as it flows, the outside of it cools in the cool air, and gets covered with slag and cinders, something like those which you may see thrown out of the furnaces in the Black Country of Staffordshire.

Sometimes these slag coverings cling together above the lava stream, and make a tunnel, through the cracks in which you may see the fiery river rushing and roaring down below. But mostly they are kept broken and apart, and roll and slide over each other on the top of the lava, crashing and clanging as they grind together with a horrid noise. Of course that stream, like all streams, runs towards the lower grounds. It slides down glens, and fills them up; down the beds of streams, driving off the water in hissing steam; and sometimes (as it did in Iceland a few years ago) falls over some cliff, turning what had been a water-fall into a fire-fall, and filling up the pool below with blocks of lava suddenly cooled, with a clang and roar like that of chains shaken or brazen vessels beaten, which is heard miles and miles away.

Of course, woe to the crops and gardens which stand in its way. It crawls over them all and eats them up. It shoves down houses; it sets woods on fire, and sends the steam and gas out of the tree-trunks hissing into the air. And (curiously enough) it does this often without touching the trees themselves. It flows round the trunks (it did so in a wood in the Sandwich Islands a few years ago), and of course sets them on fire by its heat, till nothing is left of them but blackened posts. But the moisture which comes out of the poor tree in steam blows so hard against the lava round that it can never touch the tree, and a round hole is left in the middle of the lava where the tree was. Sometimes, too, the lava will spit out liquid fire among the branches of the trees, which hangs down afterwards from them in tassels of slag, and yet (by the very same means) the steam in the branches will prevent the liquid fire burning them off, or doing anything but just scorch the bark.

But I can tell you a more curious story still. The lava stream, you must know, is continually sending out little jets of gas and steam: some of it it may have brought up from the very inside of the earth; most of it, I suspect, comes from the damp herbage and damp soil over which it runs. Be that as it may, a lava stream out of Mount Ætna, in Sicily, came once down straight upon the town of Catania. Everybody thought that the town would be swallowed up; and the poor people there (who

knew no better) began to pray to St. Agatha—a famous
saint, who, they say, was martyred there ages ago, and
who (they fancy) has power in heaven to save them from
the lava stream.   And really what happened was enough

LAVA STREAM AT CATANIA.

to make ignorant people, such as they were, think that
St. Agatha had saved them.

The lava stream came straight down upon the town
wall.   Another foot, and it would have touched it, and
have begun shoving it down with a force compared with

LAVA LANDSCAPE, ICELAND.

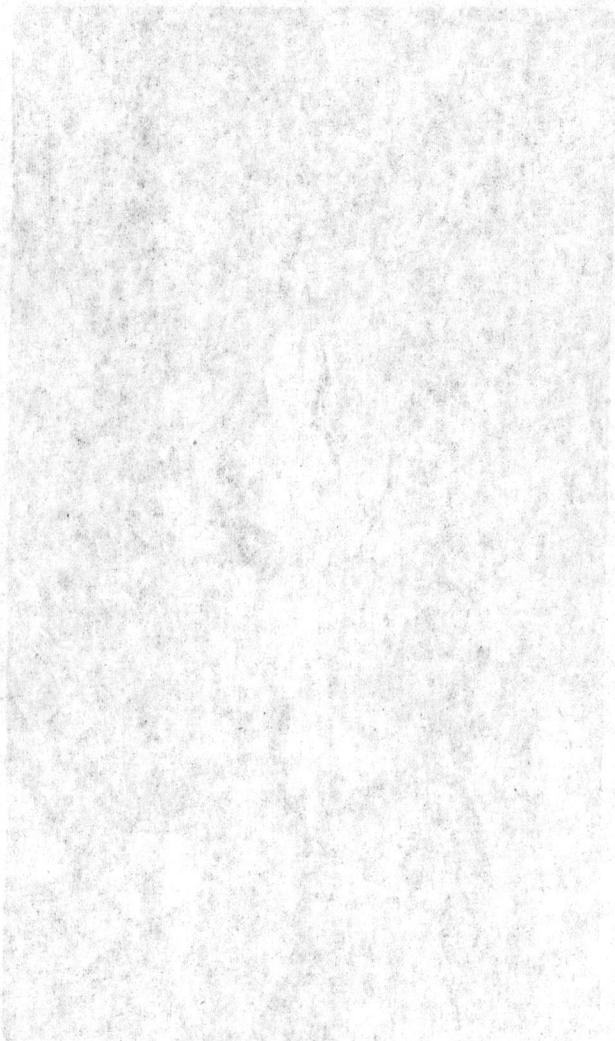

which all the battering-rams that you ever read of in
ancient histories would be child's toys. But, lo and
behold! when the lava stream got within a few inches of
the walls it stopped, and began to rear itself upright, and
build itself into a wall beside the wall. It rose and rose,
till I believe in one place it overtopped the wall, and
began to curl over in a crest. All expected that it would
fall over into the town at last; but no, there it stopped,
and cooled, and hardened, and left the town unhurt. All
the inhabitants said, of course, that St. Agatha had done
it; but learned men found out that the lava was so full of
gas, which was continually blowing out in little jets, that
when it reached the wall it actually blew itself back from
the wall, and as the wall was luckily strong enough not to
be blown down, the lava kept blowing itself back till it
had time to cool. And so there was no miracle at all in
the matter; and the poor people of Catania had to thank
not St. Agatha, and any interference of hers, but simply
Him who can preserve, just as He can destroy, by those
laws of nature which are the breath of His mouth and the
servants of His will.

But in many a case the lava does not stop. It rolls
on and on over the downs and through the valleys, till it
reaches the seashore, as it did in Hawaii, in the Sandwich Islands, in 1869. And then it cools, of course, but
not often before it has killed the fish by its sulphurous
gases and heat, perhaps for miles around. And there is
good reason to believe that the fossil fish which we so
often find in rocks, perfect in every bone, lying sometimes

in heaps, and twisted (as I have seen them) as if they had died suddenly and violently, were killed in this very way, either by heat from lava streams or else by the bursting up of gases poisoning the water, in earthquakes and eruptions in the bottom of the sea. I could tell you many stories of fish being killed in thousands by earthquakes and volcanoes during the last few years. But we have not time to tell about everything.

And now you will ask me, with more astonishment than ever, what possible use can there be in these destroying streams of fire? And certainly, if you had ever seen a lava stream even when cool, and looked down, as I have done, at the great river of rough black blocks streaming away far and wide over the land, you would think it the most hideous and the most useless thing you ever saw. But there is One who told men to judge not according to the appearance, but to judge righteous judgment. He said that about matters spiritual and human; but it is quite as true about matters natural, which also are His work, and all obey His will.

Now if you had seen, as I have seen, close round the edges of these lava streams, and sometimes actually upon them, or upon the great bed of dust and ashes which have been hurled far and wide out of ancient volcanoes, happy homesteads, rich crops, hemp and flax and wheat, tobacco, lucerne roots, and vineyards laden with white and purple grapes, you would have begun to suspect that the lava streams were not, after all, such very bad neighbours.

And when I tell you that volcanic soils (as they are

called), that is soil which has at first been lava or ashes, are generally the richest soils in the world—that, for instance, there is soil in the beautiful island of Madeira so thin that you cannot dig more than two or three inches down without coming to the solid rock of lava, or, what is harder even, obsidian (which is the black glass which volcanoes sometimes make, and which the old Mexicans used to chip into swords and arrows, because they had no steel)—and that this soil, thin as it is, is yet so fertile, that in it used to be grown the grapes of which the famous Madeira wine was made—when you remember this, and when you remember, too, the Lothians of Scotland, then you will perhaps agree with me, that there is great use in lava and ashes upon the surface of the earth.

For see—down below under the roots of the mountains, like a chemist in his laboratory, Nature works continually, melting together all the rocks, which are the bones and leavings of the old worlds. If they stayed down below there, they would be of no use, while they will be of use up here in the open air. For, year by year, by the washing of rain and rivers—and also, I am sorry to say, by the ignorant and foolish waste of mankind—thousands and millions of tons of good stuff are running into the sea every year, which would, if it could be kept on land, make food for men and animals, plants and trees. So, in order to supply the continual waste of this upper world, there is a continual melting up of the under world, and a pouring of it out by volcanoes like manure, to renew the face of the earth.

K

VOLCANOES AND EARTHQUAKES.

CHAPTER IV.

MAN-MAKERS.—(*Continued.*)

IN these lava rocks and ashes which are sent up, there
are certain substances without which men cannot live,
without which a stalk of corn or grass cannot grow.
Without potash, without magnesia—both of which are
in your veins and mine—without silicates (as they are
called), which give flint to the stems of corn and of
grass, and so make them stiff and hard, and able to
stand upright—and very probably without the carbonic
acid gas, which comes out of the volcanoes, and is taken
up by the leaves of plants, and turned by them into
solid wood—without all these things, and I suspect
without a great many more things which come out of
volcanoes, I do not see how this beautiful green world
could get on at all.

Of course, when the lava first cools on the surface of
the ground it is hard enough, and therefore barren
enough. But rain sets to work upon it at once, and
with that, as by a spade, century after century, and age
after age, the lava stream is dug down, atom by atom,
and silted over the country round in rich manure. So
that, if rough and hasty in pumping treasures up out of
her mine with her great steam-pumps, Nature shows
herself delicate and tender and kindly enough in giving
them away afterwards.

Nay, even the fine dust which is sometimes blown
out of volcanoes is useful to countries far away. So
light it is, that it rises into the sky, and is wafted by the
wind across the seas.

So, in the year 1783, ashes from the Skaptar
Jokull, in Iceland, were carried over the north of Scotland, and even into Holland, hundreds of miles to the
south.

So, again, when in the year 1812 the volcano of
St. Vincent, in the West India Islands, poured out
torrents of lava, after mighty earthquakes which shook
all that part of the world, a strange thing happened
(about which I have often heard from those who saw
it) in the island of Barbadoes, eighty miles away.

When the sun rose in the morning (it was a Sunday
morning), the sky remained more dark than any night;
and all the poor negroes crowded terrified out of their
houses into the streets, fancying the end of the world
was come. But a learned man who was there, finding

that, though the sun was risen, it was still pitchy dark, opened his window, and found that it was stuck fast by something on the ledge outside, and when he thrust it open, found the ledge covered deep in soft red dust; and he instantly said, like a wise man as he was, " The

CRACK IN LAVA FIELD.

volcano of St. Vincent must have broken out, and these are the ashes from it." Then he ran down-stairs and quieted the poor negroes, telling them not to be afraid, for the end of the world was not coming just yet. But still the dust went on falling, till the whole island, I am told, was covered an inch thick; and the same thing happened in other islands round. People thought—and they had reason to think from what had often happened elsewhere—that though the dust might hurt the crops for that year, it would make them richer in years to come, because it would act as manure upon the soil; and so it did after a few years; but it did terrible damage at the time, breaking off the boughs of trees, and covering up the crops; and in St. Vincent itself whole estates were ruined.

It was a frightful day, but I know well that behind there was a Why for its happening, and happening, too, about that very time, which all who know the history of negro slavery in the West Indies can guess for themselves; and confess, I hope, that in this case, as in all others, what seems most severe is often most just and kind.

But now I see you want to ask a question. Let us have it out. I would sooner answer one question of yours than tell you ten things without your asking.

Is there potash and magnesia and silicates in the soil here? and if there is, where did they come from? for there are no volcanoes in England.

Yes! there are such things in the soil; and little enough of them, as the farmers know too well.

But where the rich part, or staple, of our soils came from first it would be very difficult to say, so often has England been made, and unmade, and remade, and her materials afresh sifted every time. But if you go to the

Lowlands of Scotland, you may soon see where the
staple of the soil came from there, and that I was right
in saying that there were atoms of lava in every Scotch
boy's broth. Not that there were ever (as far as I know)
volcanoes in Scotland or in England. Nature has more
than one string to her bow, or two strings either. So
when she pours out her lavas, she does not always pour
them out in the open air.

VOLCANIC ARCH AT SEA.

Sometimes she pours them out at the bottom of the
sea, as she did in the north of Ireland and the south-
west of Scotland, when she made the Giant's Causeway,
and Fingal's Cave in Staffa, too, at the bottom of the
old chalk ocean, ages and ages since. Sometimes she
squirts them out between the layers of rock, or into
cracks which the earthquakes have made, in what are
called trap dykes; of which there are plenty to be seen
in Scotland, and in Wales likewise.

Then she lifts the earth up from the bottom of the sea, and sets the rain to wash away all the soft rocks, till the hard lava stands out in great hills upon the surface of the ground. Then the rain begins eating away those lava hills likewise, and manuring the earth

FINGAL'S CAVE.

with them. And wherever those lava hills stand up whether great or small, there is pretty sure to be rich land around them.

If you look at the Geological Map of England and

Ireland, and the red spots upon it, which will show you
where these old lavas are, you will see how much of
them there is in England, at the Lizard Point in Cornwall, and how much more in Scotland and the north of
Ireland. In South Devon, in Shropshire—with its beautiful Wrekin, and Caradoc, and Lawley—in Wales, round
Snowdon (where some of the soil is very rich), and
above all in the Lowlands of Scotland, you see these
red marks, showing the old lavas, which are always fertile, except the poor old granite, which is of little use
save to cut into building stone, because it is too full of
quartz—that is, flint.

Think of this the next time you go through Scotland
on the railway, especially when you get near Edinburgh.
As you run through the Lothians, with their noble crops
of corn and roots and grasses, and their great homesteads,
each with its engine chimney, which makes steam do
the work of men, you will see rising out of the plain
hills of dark rock, sometimes in single knobs like Berwick
Law or Stirling Crag, sometimes in noble ranges like
Arthur's Seat, or the Sidlaws, or the Ochils.

Think what those black bare lumps of whinstone are,
and what they do. Remember they are mines—not gold-
mines, but something richer still—food-mines—which
were thrust into the inside of the earth, ages and ages
since, as molten lava rock, and then cooled and lifted up,
and pared away with ice-plough and rain-spade, and
spread over the wide carses round, to make in that bleak
northern climate, which once carried nothing but fir-trees

VOLCANIC COAST.

and heather, a soil fit to feed a great people, to cultivate in them industry and science and valiant self-dependence and self-help, and to gather round the Heart of Mid-lothian and the Castle Rock of Edinburgh the stoutest and the ablest little nation which has been since the Greeks who fought at Salamis.

Of those Greeks you have read, or ought to read, in Mr. Cox's "Tales of the Persian War." Some day you will read of them in their own books, written in their grand old tongue. Remember that Providence made them, as she has made the Scotch, by first preparing a country for them which would call out all their courage and their skill, and then by giving them the courage and the skill to make use of the land in which she had put them.

And now think what a wonderful fairy tale you might write for yourself—and every word of it true—of the adventures of one atom of Potash or some other Salt, no bigger than a needle's point, in such a lava stream as I have been telling of. How it has run round and round, and will run round age after age, in an endless chain of change. How it began by being molten fire underground; how then it became part of a hard cold rock, lifted up into a cliff, beaten upon by rain and storm, washed down into the soil of the plain, till, perhaps, the little atom of mineral met with the rootlet of some great tree, and was taken up into its sap in spring, through tiny veins, and hardened the next year into a piece of solid wood.

And then how that tree was cut down, and its logs, it may be, burnt upon the hearth, till the little atom of mineral lay among the wood-ashes, and was shovelled out and thrown upon the field and washed into the soil again, and taken up by the roots of a clover plant, and became an atom of vegetable matter once more.

And then how, perhaps, a rabbit came by, and ate the clover, and the grain of mineral became part of the rabbit; and then how a hawk killed that rabbit, and ate it, and so the grain became part of the hawk; and how the farmer shot the hawk, and it fell perchance into a stream, and was carried down into the sea; and when its body decayed, the little grain sank through the water, and was mingled with the mud at the bottom of the sea.

But do its wanderings stop there ?

No, not so. Nothing upon this earth continues in one stay. That grain of mineral might stay at the bottom of the sea a thousand or ten thousand years, and yet the time would come when the great chemist Nature would set to work on it again. Slowly, perhaps, it would sink that mud so deep, and cover it up with so many fresh beds of mud, or sand, or lime, that under the heavy weight, and perhaps, too, under the heat of the inside of the earth, that mud would slowly change to hard slate rock; and ages after, it may be, it might melt that slate rock up once more, and blast it out; and then through the mouth of a volcano the little grain of mineral might rise into the open air again to make fresh soil, as it had done thousands of years before. For Nature can manufacture many different things out of the same materials. She may have so wrought with that grain of mineral, that she may have formed it into part of a precious stone, and men may dig it out of the rock, or pick it up in the river-bed, and polish it, and set it and wear it. Think of that—that in the jewels which your mother or your sisters wear, or in your father's signet ring, there may be atoms which were part of a live plant, or a live animal, millions of years ago, and may be parts of a live plant or a live animal millions of years hence.

Think over again, and learn by heart, the links of this endless chain of change :

Fire turned into Stone.

Stone into Soil.

Soil into Plant.

Plant into Animal.

Animal into Soil.

Soil into Stone.

Stone into Fire again.

And then Fire into Stone again, and the old thing run round once more.

So it is, and so it must be. For all things which are born in Time, must change in Time, and die in Time, till that Last Day of this our little earth, in which,

> "——like the baseless fabric of a vision,
> The cloud-capp'd towers, the gorgeous palaces,
> The solemn temples, the great globe itself,
> Yea, all which it inherit, shall dissolve;
> And, like insubstantial pageant faded,
> Leave not a rack behind."

So all things change and die, and so your body too must change and die—but not yourself. Nature made your body, and she must unmake it again, as she unmakes all her works in Time and Space; but you, your Soul, and Life, and Self, she did not make, and over you she has no power. For you were not, like your body, created in Time and Space; and you will endure though Time and Space should be no more: because you are the child of the Living God, who gives to each thing its own body, and can give you another body, even as seems good to Him.

MINES

MINES.

CHAPTER I.

A TIN-MINE.

THE tin-mines of St. Just are the most curious mines of Cornwall.

We set out with a fog so dense that we could not see ten paces before us, and were obliged to shout for fear of losing one another. At the same time the air was fresher than is usual in July, and a fine rain continually falling penetrated to the skin. This weather lasted nearly all day, but as we passed the greater part of it underground it did not inconvenience us.

On our way to the mines we passed a number of sheds and workshops for mechanical preparation of the ore. In the shops nearest of these the mineral is broken and tried, farther along it is pulverised under the mechanical pestles, then it is passed into oscillating sieves, then it

is washed on inclining wooden tables, which retain the
heaviest particles and let run away the lighter, thus
cleansing the mineral of nearly all its stone-dust.

Both men and women are employed in this work. The
men wear an apron tied to the waist, and the foreman a
cap of particular shape. The women also are dressed in

OVER A BROKEN-IN MINE.

aprons, with hoods or nets on their heads. Every one
seems healthy and lively, often singing at their work.

We entered the room of extracting machines, where
order and cleanliness also reigned, and where all the workmen attended to their duties more quietly than the women
of the neighbouring workshops. I raised my head, and on

one of the walls of the room read this inscription : " We seek hidden treasures." If mining work has ever been clearly defined, it is in this simple phrase. Hidden, indeed, are the treasures which the miner seeks. What patience, courage, money, and ardour, are needed in his bold endeavours to bring them out to light !

MINERAL TRAIN.

I must, however, pass over the numerous dangers which surround the miner. One may almost say that all the elements conspire against him : the air, which is often insufficient in quantity, becomes vitiated in the deep dark maze of caverns which surrounds him ; water on all sides threatens to inundate him ; the earth above him, by sudden and unforeseen breaking away, crushes, maims, or kills him; and gases, igniting by accident in the mine, often cause terrible explosions which blind and disfigure him.

Equally bold with the soldiers of land and sea are these obscure miners, these pioneers of industry, who have no fewer perils to encounter, and with no admiration from around to urge them on. The public, indeed, ignore the merit of these rough and valuable labourers' existence, until their attention is all at once directed towards them by some lamentable accident which buries alive all the workers of a whole mine, and plunges hundreds of families into poverty and woe.

What is true of all mines is, fortunately for the inhabitants of St. Just, less true of tin-mines than of mines of other kinds. Dangers of all kinds increase as we pass from the mineral to the coal mines, where bad air is more prevalent and gas explosions are more frequent.

One of the mines we found possessed of the title Providence Mine, a name well chosen; and, under the shield of its good name, we resolved that into this we would descend; though at the prospect held out by the glances of curiosity we cast from the top of the shaft, we confess that, in spite of its protective name, we somewhat shuddered at the thought. Before descending we learn that Providence Mine is not the least interesting of the district. It produces about thirty tons of tin ore per month, employs three hundred workmen, and is explored about six hundred feet below the level of the sea.

TIN-MINE AT ST. JUST.

"Under the sea!" you exclaim. "Yes, certainly; and if you will descend into it you will soon have the most convincing proof of it."

Before leading us into the interior of his mine, the foreman, in whom we found an agreeable guide, made us dress

WOMAN OF THE MINE.

in the mining costume—shirt, jacket, and white flannel pantaloons, huge leather boots, black felt hat of a low round shape and hard as stone, with a lamp fastened in the front of it. Thus equipped for the descent, we certainly presented to each other a most remarkable appear-

ance. And this costume is a necessary precaution, so much
so that even the Prince of Wales himself, when he visited
the mines, and surveyed the subterranean works round St.
Just, was equipped, like us, in the flannel and hat and
lamp of the miner.

When we thus prepared, we were asked if we wished
to descend by the ladders or the man-engine.

" Captain," said I, " what is the depth of the shaft ? "

" Two hundred fathoms."

" Very well, we will go down by the man-engine," for
to descend a ladder to the depth of two hundred fathoms
seemed to me at least to be an impossibility. This
machine I cannot better represent than by an enormous
oscillating pole, inclined or vertical, according as the axe
of the shaft is itself inclined or vertical, and provided at
certain distances with a little stage or shelf, on each of
which a person stands leaning against the pole. It is a
sort of movable ladder, the stages representing the rounds,
the distances between each being more than the height of
a man. The groove in which this works extends the
whole depth of the shaft, fastened at the top to the beam
of the engine which works it. It is what may fairly be
called a multiplication of " lifts." These man-engines
were invented to spare the workman the daily fatigue of
a descent and ascent by fixed ladders of many hundreds
of yards in length. This exercise, repeated twice a day,
besides causing the loss of much precious time, quickly
produces amongst the labourers diseases which lead them
step by step to their graves. Many of them are of a

much more elegant construction than the one I have described : the stages are wide and comfortable, with a balustrade on each side.

This is how the apparatus is contrived. A movement of the steam-engine beam causes this man-engine to descend about five feet. The workman who has descended in the engine, passes immediately that it stops on to a little ledge, leaning against the partition wall of the pit. A second movement of the balance causes the engine to re-

MINER'S HAT.

mount, bringing its next lower stage up to the level of the man waiting on the little ledge ; the man passes on to the man-engine without trouble or hesitation, which immediately descends, bringing him to a next lower ledge, on to which he steps. The machine and man repeat the same operation till the man reaches his destination.

Lastly, I must not forget to state that the machine stops for a very short time after each movement, but just sufficient to allow the men to pass from the ledge to the machine, and *vice versâ*.

An example will serve to make you more fully comprehend the utility of the man-engine.

Suppose that, at each oscillation of the steam-engine which works it, it descends five feet, and that the machine rises and falls twenty times per minute, that is to say, performs ten movements in one direction and ten in the opposite direction, that would enable its user to descend, if descending, or to ascend, if ascending, at the rate of fifty feet per minute, or fifteen hundred feet in half an hour. It would take double this time for the workmen to perform the same distance by the ladders, besides incurring an immense amount of fatigue. It would take much less time, it is true, to go at the end of the cable in the basket; but the shaft is nearly always otherwise occupied, besides which this course is highly dangerous, for sometimes the cable breaks, and hurls the basket and contents to the bottom of the mine.

Partly to avoid the fatigue of the fixed ladders, long as that of Jacob, and partly to try such a curious apparatus, we had decided for the man-engine.

Our guide went first, my bolder friend second, I third. My timid friend, afraid at the sight of the sea, and not wishing to run the risk of a journey to the bottom of these dark caves, stayed outside.

At the usual signal, " Go ahead," the machine gives way, and we are in motion.

At first all goes well. One is struck with surprise at

the gigantic balancing by which he is moved away. But
darkness gathers, and the lamp fixed in the hat-brim
gives very little light. So you hesitate to pass from the
stage of the apparatus to the ledge in the shaft; you can
scarcely stand firm on these yielding boards; but go on
to them at once you must, for the machine does not wait.

As I obeyed to the best of my ability this mechanical
rhythm, passing alternately from the machine to the
shaft, and from the shaft to the machine, avoiding as
much as possible the workmen, a detachment of whom
was remounting whilst we descended, I heard all at once
a cry. I looked down below. It was from my companion, who, finding the plank on which he had set his
foot occupied by one of the men who were ascending,
had not been quick enough to pass to the opposite side,
and was making a disturbance instead of waiting quietly
for another pulsation of the man-engine. In remounting,
the machine had caught him; fortunately, he was holding firmly to the iron rail which is put about the height
of the arm; but I saw one moment his body balanced in
the air. I heard the captain cry, and thought all was
over.

Happily, my friend did not lose his presence of mind.
The order to stop the machine was immediately given,
and the guide, who had already laid hold of my friend to
hold him, drew him up safe and sound on the bench.
My friend recovered his calmness, and we were drawn up
to the daylight.

When the first emotion was over, I persisted in

descending again by the man-engine with our guide. The latter bluntly refused. He said it was not altogether an unheard-of thing for a man caught in the apparatus to lose an arm or leg, and some had even been nearly killed by the stroke; he continued, that our lives had been confided to his care, and that if we did not wish to repeat an accident the result of which might have been fatal, we must descend by the ladders.

I submitted myself to these reasons, and then commenced the longest, the most laborious descent I ever made in my life. Step after step, ladder after ladder, and it was only at very long intervals we found a landing or anything on which we might sit down and recover our breath.

Sometimes I reflected that this descent must be followed by an ascent as long, but far more difficult. This idea almost made my hair stand on end. Here and there we reached openings of galleries, but when I wished to explore them, "Lower, lower!" cried the guide; "the works are left off here, there is nothing to see;" and we again began to descend. At last we arrived, not at the bottom of the mine (it would have taken an hour to reach that) but half-way, where we made a final halt.

I heard the miners hammers resounding on the rock, and gave thanks to Heaven.

The place where they worked was a shaft descending from one of the side galleries—a little mine within a mine. We walked along the gallery to the mouth of the shaft, and then, attached to the end of a strong rope unrolled from a windlass, we descended to this cavern of treasure. Here we found a group of miners occupied in chipping away with chisel and hammer quartz rock or unstratified rock. It was hard work. The crystal and

MINERS AT WORK.

metal mixed with it glistened in the faint light of candles, and sparks flew under each blow of the steel.

In a neighbouring gallery, where the work was on the floor, there were three men engaged at one chisel: one was sitting on the ground, and he held the chisel between

his hands, the two others, standing with immense hammers, struck by turns the chisel-head, like two black-
smiths striking on the anvil. The mineral came away
in wedges and flakes, was then put into small waggons,
and conveyed away on a railroad as far as the foot of the
shaft, to be hoisted to the surface.

After watching for a time, we ascended one part of
the ladders, then passed through a long tunnel where the
water of the pumps emptied itself running like a little
brook.

"Taste this water," said the captain to me. I
stooped, and, like Diogenes, taking a little water in the
hollow of my hand, put it to my lips. I hastily threw it
down; it was bitter as magnesia, as a solution of Epsom
salts. It was sea-water, rendered still more unpleasant
to the taste by the metals of the rocks through which it
had percolated. Added to this was a taste of ink, and a
sediment of red ochre marked the course of the water.

At this point there could be no doubt that the galleries
we were exploring were under the ocean, for its water
rained down upon us through thin roofs, and it was not
difficult to imagine that we heard the beating of its surf
above. Continuing our course in the same tunnel, we
were astonished as soon as we reached daylight to find
ourselves at the sea-coast. We thanked Providence, the
protector of the mine, for having permitted us to make
so interesting an excursion without hindrance or harm.
We soon forgot the discomforts, but never the instruction,
of our subterranean tour.

MINES.

CHAPTER II.

A COAL-MINE, AS IT WAS.

LET me tell you now something of a geological pro-
duct, happily well known: I mean coal.

Coal-beds, as of course you know, are commonly said
to be composed of vegetable matter, of the leaves and
stems of ancient plants and trees. How was this startling
fact—for fact it is—discovered?

You will have a right to say at first starting, "Coal
is utterly different in look from leaves and stems. The
only property which they seem to have in common is
that they can both burn."

True. But difference of mere look may be only owing
to a transformation, or series of transformations. There
are plenty in nature quite as great, and greater. What

can be more different in look, for instance, than a green field of wheat and a basket of loaves at the baker's? And yet there is, I trust, no doubt whatsoever that the bread has been once green wheat, and that the green wheat has been transformed into bread.

But you may say, "Yes, but we can see the wheat growing, flowering, ripening, reaped, ground, kneaded, baked. We see, in the case of bread, the processes of the transformation going on; but in the case of coal we do not see the wood and leaves being actually transformed into coal, or anything like it."

But suppose we laid out the wheat on a table in a regular series, such as you may see in many exhibitions of manufactures, beginning with the wheat plant at one end, and ending with the loaf at the other, and called in to look at them a savage who knew nothing of agriculture and nothing of cookery—called in, as an extreme case, the man in the moon, who certainly can know nothing of either; for as there is neither air nor water round the moon, there can be nothing to grow there, and therefore nothing to cook—and suppose we asked him to study the series from end to end.

Do you not think that the man in the moon, if he were half as shrewd as Crofton Croker makes him in his conversation with Daniel O'Rourke, would answer after a due meditation, "How the wheat plant got changed into the loaf I cannot see from my experience in the moon; but that it has been changed, and that the two are the same thing I do see, for I see all the different

COAL-FOREST.

stages of the change?" And so I think you may say of the wood and the coal.

The man in the moon would be quite reasonable in his conclusion; for it is a law, a rule, and one which you will have to apply again and again in the study of natural objects, that however different two objects may look in some respects, yet if you can find a regular series of gradations between them, with all shades of likeness, first to one of them and then to the other, then you have a fair right to suppose them to be only varieties of the same species, the same kind of thing, and that, therefore, they have a common origin.

Now apply the same reasoning to coal. You may find about the world—you may see even in England alone—every gradation between coal and growing forest. You may see the forest growing in its bed of vegetable mould; you may see the forest dead and converted into peat, with stems and roots in it; that, again, into sunken forests, like those to be seen below high-water mark on almost every coast on this island. Then you find gradations between them and beds of lignite, or wood coal; then gradations between lignite and common or bituminous coal; and then gradations between common coal and culm, or anthracite, such as is found in South Wales. Have you not a right to say, "These are all but varieties of the same kind of thing—namely, vegetable matter? They have a common origin—namely, woody fibre? And coal, or rather culm, is the last link in a series of transformations from growing vegetation."

This is our first theory. Let us try to verify it, as scientific men are in the habit of doing, by saying, If that be true, then something else is likely to be true too.

If coal has all been vegetable soil, then it is likely that some of it has not been quite converted into shapeless coal. It is likely that there will be vegetable fibre still to be seen here and there; perhaps leaves, perhaps even stems of trees, as in a peat bog. Let us look for them.

You will not need to look far. The coal, and the sands and shales which accompany the coal, are so full of plant-remains, that three hundred species were known to Adolphe Brongniart as early as 1849, and that number has largely increased since.

Now one point is specially noticeable about these plants of the coal; namely, that they may at least have grown in swamps.

First, you will be interested if you study the coal flora, with the abundance, beauty, and variety of the ferns. Now ferns in these islands grow principally in rocky woods, because there, beside the moisture, they get from decaying vegetable or decaying rock, especially limestone, the carbonic acid which is their special food, and which they do not get on our dry pastures, and still less in our cultivated fields. But in these islands there are two noble species, at least, which are true swamp-ferns; the Lastræa Thelypteris, which of old filled the fens, but is now all but extinct; and the Osmunda, or King fern, which (as all know) will grow anywhere if there is damp enough about the roots.   In Hampshire, in Devon, and Cornwall, and in the south-west of Ireland, it, too, is a true swampy fern.   But in the Tropics I have seen more than once noble tree-ferns growing in wet savannahs at the sea-level, as freely as in the mountain-woods ; ferns with such a stem as some of the coal ferns had, some fifteen feet in height, under which, as one rode on horseback, one saw the blazing blue sky, as through a parasol of delicate lace, as men might have long ages since have seen it, through the plumed fronds of the ferns in the coal, had there only been a man then created to enjoy its beauty.

Next we find plants called by geologists Calamites.   There is no doubt now that they are of the

OSMUNDA.

same family as our Equiseta, or horse-tails, a race which

has, over most parts of the globe, dwindled down now from twenty or thirty feet in height, as they were in the old coal measures, to paltry little weeds. The tallest Equisetum in England—the beautiful E. Telmateia—is seldom five feet high. But they, too, are mostly mud and swamp plants ; and so may the Calamites have been.

The Lepidodendrons, again, are without doubt the splendid old representatives of a family now dwindled down to such creeping things as our club-mosses, or Lycopodiums.

Now it is a certain fact, which can be proved by the

FOSSIL FERN.

microscope, that a very great part of the best coal is actually made up of millions of the minute seeds of club-mosses, such as grow—a few of them, and those very small—on our moors ; a proof, surely, not only of the vast amount of the vegetation in the coal-making age, but also of the vast time during which it lasted. The Lepidodendra may have been fifty or sixty feet high. There is not a Lycopodium in the world now, I believe, five feet high. But the club-mosses are now, in these islands and elsewhere, lovers of wet and peaty soils, and so may their huger prototypes have been, in the old forests of the coal.

Of the Sigillariæ we cannot say as much with certainty, for botanists are not agreed as to what low order of

FOSSIL FERN.

flowerless plants they belong. But that they rooted in flat clay beds there is proof, as you will hear presently.

And as to the Conifers, or pine-like trees; the Dadoxylon, of which the pith goes by the name of Stenbergia, and the uncertain tree which furnishes in

some coal-measures bushels of a seed connected with
that of the yew, we may suppose that they would
find no more difficulty in growing in swamps than the
cypress, which forms so large a portion of the vegetation in the swamps of the Southern United States.

I have given you these hints, because you will naturally
wish to know what sort of a world it was in which all
these strange plants grew and turned into coal.

My answer is, that it was most probably just like the
world in which we are living now, with the one exception
that the plants and animals are different.

It was the fashion a few years since to explain the coal
—like other phenomena of geology—by some mere hypothesis of a state of things quite unlike what we see now.
We were brought up to believe that in the Carboniferous,
or coal-bearing era, the atmosphere was intensely moist
and hot, and overcharged with carbonic acid, which had
been poured out from the interior of the planet by volcanic eruptions, or by some other convulsions. I forget
most of it now : and really there is no need to remember ;
for it is all, I verily believe, a dream—an attempt to
explain the unknown not by the known, but by the still
more unknown. You may find such theories lingering
still in sensational school-books, if you like to be unscientific. If you prefer, on the other hand, to be scientific,
you will listen to those who tell you that instead of there
having been one unique carboniferous epoch, with a
peculiar coal-making climate, all epochs are carboniferous
if they get the chance ; that coal is of every age, from

that of the Scotch and English beds, up to the present day.

Coal is forming now, I doubt not, in many places on the earth, and would form in many more, if man did not interfere with the processes of wild nature, by draining the fens and embanking the rivers.

Let me by a few words prove this statement.

When it was first seen that coal had been once vegetable, the question arose—How did all these huge masses of vegetable matter get there? The Yorkshire and Derbyshire coal-fields, I hear, cover seven or eight hundred square miles; the Lancashire is about two hundred. How large the North Wales and the Scotch fields are, I cannot say. But doubtless a great deal more coal than can be got at lies under the sea, especially in the north of Wales. Coal probably exists over vast sheets of England and France, buried so deeply under later rocks, that it cannot be reached by mining.

As an instance, a distinguished geologist has long held that there are beds of coal under London itself, which rise, owing to a peculiar disturbance of the strata, to within a thousand or twelve hundred feet of the surface. and that we or our children may yet see coal-mines in the marshes of the Thames.

And more, it is a probable fact that only a portion of the coal-measures is left. A great part of Ireland must once have been covered with coal, which is now destroyed. Indeed, it is likely that the coal now known of in Europe and America is but a remnant of what existed there in former ages, and has been eaten away by the inroads of the sea.

Now, whence did all that enormous heap of vegetable soil come?

Off some neighbouring land, was the first and most natural answer. It was a rational one. It proceeded from the known to the unknown.

It was clear that these plants had grown on land; for they were land-plants.

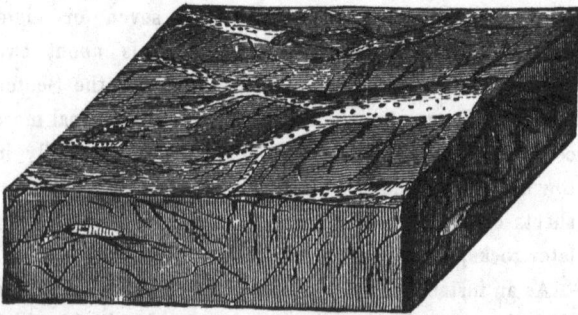

STIGMARIA FOSSIL, ON BLOCK OF COAL.

It was clear that there must have been land close by, for between the beds of coal, as you all know, the rock is principally coarse sandstone, which could only have been laid down in very shallow water.

It was natural, then, to suppose that these plants and trees had been swept down by rivers into the sea, as the sands and muds which buried them had been. And it was known that at the mouths of certain rivers—the

Mississippi, for instance—vast rafts of dead floating trees accumulated ; and that the bottoms of the rivers were often full of snags, &c.; trees which had grounded, and stuck in the mud ; and why should not the coal have been formed in the same way ?

Because—and this was a serious objection—then surely

SIGILLARIA FOSSIL, STANDING IN COAL BED.

the coal would be impure—mixed up with mud and sand till it was not worth burning. Instead of which, the coal is pure vegetable, parted sharply from the sandstone which lies on it.

The only other explanation was, that the coal vegetation had grown in the very places where it was found.

But that seemed too strange to be true, till that great geologist, Sir W. Logan—who has since done such good work in Canada—showed that every bed of coal had a bed of clay under it, and that that clay always contained fossils called Stigmaria. Then it came out that the Stigmaria in the under clay had long filaments attached to them, while, when found in the sandstones or shales, they had lost their filaments, and seemed more or less rolled— in fact, that the natural place of the Stigmaria was in the under clay:

Then Mr. Binney discovered a tree—a Sigillaria, standing upright in the coal-measures with its roots attached. Those roots penetrated into the under-clay of the coal, and those roots were Stigmarias. That seems to have settled the question. The Sigillarias, at least, had grown in the spot where they were found, and the clay beneath the coal-beds was the original soil on which they had grown. Just so, if you will look at any peat bog, you will find it bottomed by clay, which clay is pierced everywhere by the roots of the moss forming the peat, or of the trees, birches, alders, poplars, willows, which grow in the bog.

So the proof seemed complete, that the coal had been formed out of vegetation growing where it was buried.

I have dwelt at length on these further arguments, because they seem to me as pretty a specimen as I can give my readers of that regular and gradual induction, that common sense regulated, by which geological theories are worked out.

QUARRIES.

CHAPTER III.

SLATE.

SLATES come from the hills of Cumberland, Westmoreland, and North Wales. How beautiful and how curious an object is a common slate!

Beautiful: not only for the compactness and delicacy of its texture, and for the regularity and smoothness of its surface, but still more for its colour.

Beautiful is the slate, and curious likewise, venerable; a most ancient and elaborate work of God. And yet it was at first nought but an ugly lump of soft and shapeless ooze, for slate is hardened mud.

That slate may have been once mud, is made probable by the simple fact that it can be turned into mud again. If you grind up slate, and then analyze it, you will find

that when a tree dies its wood decays first, its bark last ?
It is so, especially in the Tropics. There one may see
huge dead trees with their bark seemingly sound, and
their inside a mere cavern with touchwood at the bottom;
into which caverns one used to peep with some caution.
For though one might have found inside only a pair of
toucans, or parrots, or a whole party of jolly little monkeys, one was quite as likely to find a poisonous snake
four or five feet long, whose bite would have certainly
prevented me having the pleasure of writing this paper.

But is it not plain that if such trees as that sunk, their
bark would be turned into lignite, and at last into coal,
while their insides would naturally be silted up with mud
and sand ? Thus a core or pillar of hard sandstone would
be formed, which might do to the collier of the future
what they are too apt to do in the Newcastle and Bristol
collieries. For there, when the coal is worked out below,
the sandstone stems—" coal-pipes," as the colliers call
them—in the roof of the seam, having no branches, and
nothing to hold them up but their friable bark of coal, are
but too apt to drop out suddenly, killing or wounding the
hapless men below.

Or again, if we find—as we very often find—as was
found at Parkfield Colliery, near Wolverhampton, in the
year 1844—a quarter of an acre of coal-seam filled with
stumps of trees as they grew, their trunks broken off and
lying in every direction, turned into coal, and flattened as
coal-fossils so often are, by the weight of the rock above
—should we not have a right to say—These trees were

snapped off where they grew by some violent convulsion
—a storm, or a sudden inrush of water, owing to a
sudden sinking of the land, or by the very earthquake
shock itself which sank the land?

But what evidence have we of such sinkings?

The plain fact that you have coal-seam above coal-
seam, each with its bed of under-clay, and that the land
must have sunk ere the next bed of soil could have been
deposited, and the next forest have grown on it.

In one of the Rocky Mountain coal-fields there are more
than thirty seams of coal, each with its under-clay below
it. What can that mean but thirty or more subsidences
of the land, and the peat of thirty or more forests or peat-
mosses, one above the other? And now, if any reader
shall say, Subsidence? What is this quite new element
which you have brought into your argument? You told
us that you would reason from the known to the unknown.
What do we know of subsidence? You offered to explain
the thing which had gone on once by that which is going
on now. Where is subsidence going on now upon the
surface of our planet? And where, too, upheaval, such
as would bring us these buried forests up again from
under the sea-level, and make them, like our British coal-
field, dry land once more?

The answer is—Subsidence and elevation of the land
are common now, probably just as common as they were
in any age of this planet's history.

To give two instances, made now notorious by the
writings of Sir C. Lyell and other geologists. As lately

as 1819 a single earthquake shock in Cutch, at the mouth
of the Indus, sunk a tract of land larger than the Lake of
Geneva in some places to a depth of eighteen feet, and
converted it into an inland sea. The same shock raised,
a few miles off, a corresponding sheet of land some fifty
miles in length, and in some parts sixteen miles broad,
ten feet above the level of the alluvial plain, and left it
to be named by the country people the " Ullah Bund," or
Bank of God, to distinguish it from the artificial banks
in the neighbourhood.

Again, in the valley of the Mississippi, a tract which is
now, it would seem, in much the same state as central
England was while our coal-fields were being laid down,
the earthquakes of 1811—12 caused large lakes to appear
suddenly in many parts of the district, amid the dense
forests of cypresses. One of these, the " Sunk Country,"
near New Madrid, is between seventy and eighty miles in
length, and thirty miles in breadth, and throughout it, as
late as 1846, '' dead trees were conspicuous, some erect
in the water, others fallen, and strewed in dense masses
over the bottom, in the shallows, and near the shore."

If, therefore, the reader wishes to picture to himself the
scenery of what is now central England, during the period
when our coal was being laid down, he has only, I believe,
to transport himself in fancy to any great alluvial delta,
in a moist and warm climate, favourable to the growth of
vegetation. He has only to conceive wooded marshes, at
the mouth of great rivers, slowly sinking beneath the sea ;
the forests in them killed by the water, and then covered

up by layers of sand, brought down from inland, till that new layer became dry land, to carry a fresh crop of vegetation. He has thus all that he needs to explain how coal-measures were formed.

I myself saw once a scene of that kind, which I should be sorry to forget; for there was, as I conceived, coal making, or getting ready to be made, before my eyes: a sheet of swamp, sinking slowly into the sea; for there stood trees still rooted below high-water mark, and killed by the waves; while inland huge trees stood dying, or dead, from the water at their roots. But what a scene— a labyrinth of narrow creeks, so narrow that a canoe could not pass up, haunted with alligators and boa-constrictors, parrots and white herons, amid an inextricable confusion of vegetable mud, roots of the alder-like mangroves, and tangled creepers hanging from tree to tree, and overhead huge fan-palms, delighting in the moisture, mingled with still huger broad-leaved trees in every stage of decay. The drowned vegetable soil of ages beneath me; above my head, for a hundred feet, a mass of stems and boughs, and leaves and flowers, compared with which the richest hothouse in England was poor and small.

But if the sinking process which was going on continued a few hundred years, all that huge mass of wood and leaf would be sunk beneath the swamp, and covered up in mud washed down from the mountains, and sand driven in from the sea—to form a bed, many feet thick, of what would be first peat, then lignite, and last, it may be, coal, with the stems of killed trees standing up out of it into the new

N

mud and sand-beds above it, just as the Sigillaria and
other stems stand up in the coal-beds both of Britain and
of Nova Scotia, while over it a fresh forest would grow

LEPIDODENDRA.

up, to suffer the same fate—if the sinking process went on
—as that which had preceded it.

That was a sight not easily to be forgotten. But we need not have gone so far from home, at least, a few hundred years ago, to see an exactly similar one. The fens of Norfolk and Cambridgeshire, before the rivers were embanked, the water pumped off, the forests felled, and the reed-beds ploughed up, were exactly in the same state.

The vast deposits of peat between Cambridge and the sea, often filled with timber trees, either fallen or upright as they grew, and often mixed with beds of sand or mud, brought down in floods, were formed in exactly the same way; and if they had remained undrained, then that slow sinking, which geologists say is going on over the whole area of the Fens, would have brought them gradually, but surely, below the sea-level, to be covered up by new forests, and converted in due time into coal. And future geologists would have found—they may find yet, if, which God forbid! England should become barbarous and the trees be thrown out of cultivation—instead of fossil Lepidodendra and Sigillariæ, Calamites and ferns, fossil ashes and oaks, alders and poplars, bulrushes and reeds. Almost the only fossil fern would be that tall and beautiful Lastræa Thelypteris, once so abundant, now all but destroyed by drainage and the plough.

We need not, therefore, fancy any extraordinary state of things on this planet while our English coal was being formed. The climate of the northern hemisphere—Britain, at least, and Nova Scotia—was warmer than now, to judge from the abundance of ferns, and especially of tree-ferns; but not so warm, to judge from the presence of conifers (trees of the pine tribe), as the

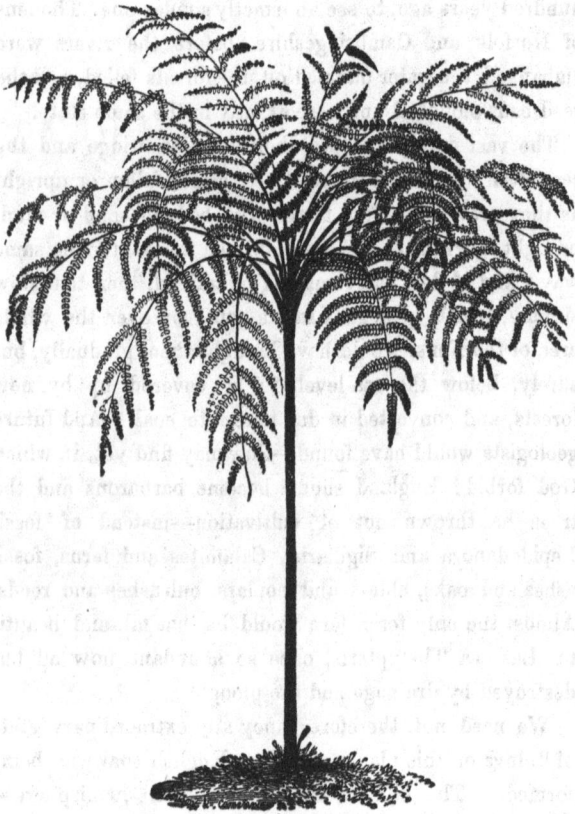

TREE FERN.

Tropics. Moreover, there must have been, it seems to

me, a great scarcity of animal life. Insects are found, beautifully preserved; a few reptiles, too, and land-shells; but how few! But, be that as it may, the fact that coal is buried forest is not affected.

Meanwhile, the shape and arrangements of sea and land must have been utterly different from what they are now. Where was that great land, off which great rivers ran to deposit our coal-measures in their deltas?

It has been supposed, for good reasons, that north-western France, Belgium, Holland, and Germany were then under the sea; that Denmark and Norway were joined to Scotland by a continent, a tongue of which ran across the centre of England, and into Ireland, dividing the northern and southern coal-fields. But how far to the west and north did that old continent stretch? Did it, as it almost certainly did long ages afterwards, join Greenland and North America with Scotland and Nor-way? Were the northern fields of Nova Scotia, which are of the same geological age as our own, and containing the same plants, laid down by rivers which ran off the same continent as ours? Who can tell now? That old land, and all record of it, save what these fragmentary coal-measures can give, are buried in the dark abyss of countless ages; and we can only look back with awe, and comfort ourselves with the thought—Let Time be ever so vast, yet Time is not Eternity.

One word more. If my readers have granted that all for which I have argued is probable, they will still have a right to ask for further proof.

They will be justified in saying, " You say that coal is transformed vegetable matter ; but can you show us how the transformation takes place ? Is it possible, according to known natural laws ? "

The chemist must answer that. And he tells us that wood can become lignite, or wood-coal, by parting with its oxygen, in the shape of carbonic acid gas, or choke-damp ; and then common, or bituminous coal, by parting with its hydrogen, chiefly in the form of carburetted hydrogen—the gas with which we light our streets. That is about as much as the unscientific reader need know. But it is a fresh corroboration of the theory that coal has been once vegetable fibre, for it shows how vegetable fibre can, by the laws of nature, become coal. And it certainly helps us to believe that a thing has been done if we are shown that it can be done.

This fact explains, also, why in mines of wood-coal carbonic acid, *i.e.* choke-damp, alone is given off. For in the wood-coal a great deal of the hydrogen still remains. But in mines of true coal, not only is choke-damp given off, but that more terrible pest of the miners, fire-damp. Now the occurrence of that fire-damp in mines proves that changes are still going on in the coal ; that it is getting rid of its hydrogen, and so progressing toward the state of anthracite or culm—stone-coal, as it is some-times called.

And is a further transformation possible ?

Yes ; and more than one.

If we conceive the anthracite cleared of all but its last

atoms of oxygen, hydrogen, and nitrogen, till it has become all but pure carbon, it would become—as it has become in certain rocks of immense antiquity—graphite, what we miscall blacklead.

And, after that, it might go through one transformation more, and that the most startling of all. It would need only perfect purification and crystallisation to become—a diamond; nothing less.

We may consider the coal upon the fire as the middle term of a series, of which

The first is live wood,

And the last, diamond.

And we may indulge safely in the fancy that every diamond in the world has probably, at some remote epoch, formed part of a growing plant.

A strange transformation; which will look to us more strange, more truly poetical, the more steadily we consider it.

The coal on the fire; the table at which I write—what are they made of?

Gas and sunbeams; with a little percentage of ash, or earthy salts, which need hardly be taken into account.

Gas and sunbeams! Yes, strange, but true.

The life of the growing plant—and what that life is who can tell?—laid hold of the gases in the air and in the soil; of the carbonic acid, the atmospheric air, the water, for that too is gas. It drank them in through its rootlets: it breathed them in through its leaf-pores, that it might distil them into sap. and bud, and leaf, and wood.

But it had to take in another element, without which the distillation and the shaping could never have taken place. It .had to drink in the sunbeams—that mysterious and complex force which is for ever pouring from the sun, and making itself partly palpable to our senses as heat and light. So the life of the plant seized the sunbeams, and absorbed them, buried them in itself— no longer as light and heat, but as invisible chemical force locked up for ages in that woody fibre.

So it is. Lord Lytton told us long ago, in a beautiful song, how

"The Wind and the Beam loved the Rose."

But Nature's poetry was more beautiful than man's. The wind and the beam loved the rose so well that they made the rose—or rather the rose took the wind and the beam, and built up out of them, by her own inner life, her exquisite texture, hue, and fragrance.

What next? The rose dies; the timber tree dies, decays down into vegetable fibre—is buried and turned to coal; but the plant cannot altogether undo its own work. Even in death and decay it cannot set free the sunbeams imprisoned in its tissue. The sun-force must stay, shut up, age after age, invisible, but strong; working at its own prison-cells; transmuting them, or making them capable of being transmuted by man, into the manifold products of coal—coke, petroleum, mineral pitch, gases, coal-tar, benzole, delicate aniline dyes, and what not, till its day of deliverance comes.

Man digs it, throws it on the fire, a black, dead-seeming lump. A corner, an atom of it, warms till it reaches the igniting point; the temperature at which it is able to combine with oxygen.

And then, like a dormant live thing, awaking after ages to the sense of its own powers, in its own needs, the whole lump is seized, atom after atom, with an infectious hunger for that oxygen which it lost centuries since in the bosom of the earth. It drinks the oxygen in at every pore, and burns.

And so the spell of ages is broken. The sun-force bursts its prison-cells and blazes into the free atmosphere, as light and heat once more; returning in a moment into the same forms in which it entered the growing leaf a thousand centuries since.

MINES.

CHAPTER IV.

A COAL-MINE, AS IT IS.

LET us now see a town in the coal-forest.

Town! you say.

Yes, indeed, for towns there are in these buried forests; and busy towns too, full of living men and boys having their tales of joy and sorrow, of danger and accident, of cowardice and bravery.

We shall best know what a coal-mine is like, for that is what we mean by a town in the coal-forests, if we try to fancy something. But first let me tell you that the entrance to a coal-pit is by what is called a shaft, which is a deep round hole in the ground like a very deep well, for the coal-forest is now a long way down under the ground. Down these deep holes the coal-workers go, and at the bottom they come to the coal-town. We will go down with some of them at another time and have a walk in one, but not until we have tried to fancy what they are like.

Well, to begin. There are two shafts or ways into the town, one at each end. So let us conceive the height of St. Paul's and of the Monument of London to represent the two shafts, and ourselves and the streets we traverse the workmen and the interior of the pit. St. Paul's Churchyard and the Monument-yard shall be the busy spaces at the bottom of the shafts; Cheapside, Cannon Street and Thames Street, shall be the main streets of the pit; the numerous streets branching off right and left from it, and Bread Street, Queen Street, &c., may be the side streets. The streets crossing these again give us an idea of the narrow continuous courts and smaller passages. The blocks of warehouses on every side of the streets would be the coal walls in the pit; and could we but suddenly let down a thick covering over our heads just a few feet above them to exclude light and air, we should have the counterpart of a northern coal-pit.

Now for what we should meet and find in the streets. Waggons and horses, and men and boys busy with hammers and picks and spades, filling and pulling waggons of coals.

But we should find the busy town by no means a busy *little* town, for it is a very large one. The distance from St. Paul's to the Monument conveys no adequate notion of the distance of a large coal-mine. We ourselves have paced along some four miles underground in direct lines,

and some five miles in winding and intersecting passages.
Yet, as we have been informed, we had but entered into
one or two of the many streets of the city, had but passed

COAL-TOWN GATE.

through the great main thoroughfares and a few side
passages, leaving miles upon miles of close and damp
passages and courts untrodden.

Tracing our progress afterwards upon a colliery map, we have come to understand how little we had explored; and there is one old pit, not many miles from Newcastle, in which, to pass through all the passages, we should be required to travel more than fifty miles. It is said that the pits at Killingworth have more than a hundred miles of streets.

It must be remembered that the whole of these streets, courts, and passages are excavations which have been and are being made by human hands. And it is said that we could reckon the entire aggregate of coal-town streets in the north of England much exceeds *one thousand* miles.

Twenty or thirty of the older and larger collieries could make up this amount of mileage, and if we should add together all the excavations in the vicinity of the Tyne, the Wear, and the Tees, we should probably be quite safe in assuming the subterranean streets and ways as extending in all over *fifteen hundred* miles; that is, by piecing together every hollow way from first to last, and from Tyne to Tees.

Think, then, of these fifteen hundred miles as the handiwork of thousands of human beings who have wrought in darkness, not because they loved the darkness rather than the light, and not because their deeds were evil, but simply because there alone, far from sunbeam and flowers and glittering streams, could they earn their honest wages, and hardly win the bread that should support home and family.

In the counties of Durham. Northumberland. and

Cumberland alone there are some thirty or forty thousand of our fellow-creatures living down in these coal-forest towns—at least working there.

But how do the people live?

When we have shut two or three hundred men and boys seven hundred or a thousand feet underground, and sometimes fourteen or fifteen hundred feet, we have sent them away from the prime necessity of life—pure atmospheric air. The farther they go along the streets of the pit, the farther they go from their natural support, and unless we can send that after them and to them, all attempts to get coal are hopeless. They must have air, or die.

Well, first you must know, that there is a *natural ventilation* produced by the mere communication of two shafts; the air will descend one and ascend the other, or partly descend and ascend both. Yet in all natural ventilation the air will take the shortest course from shaft to shaft, and therefore only ventilate a passage or two of the mine, and even those inefficiently. But the men are distributed over most of the passages, and as the pit enlarges, they proceed to greater and greater distances from the shafts and the main-way or Cheapside of the underground city. No air, of purity or amount, will reach them there, unless compelled to go to them. The interior recesses of an imperfectly ventilated pit is, as we can testify from painful recollections, one of the most pestiferous places conceivable. No tongue will ever tell what terrible sufferings have been endured in

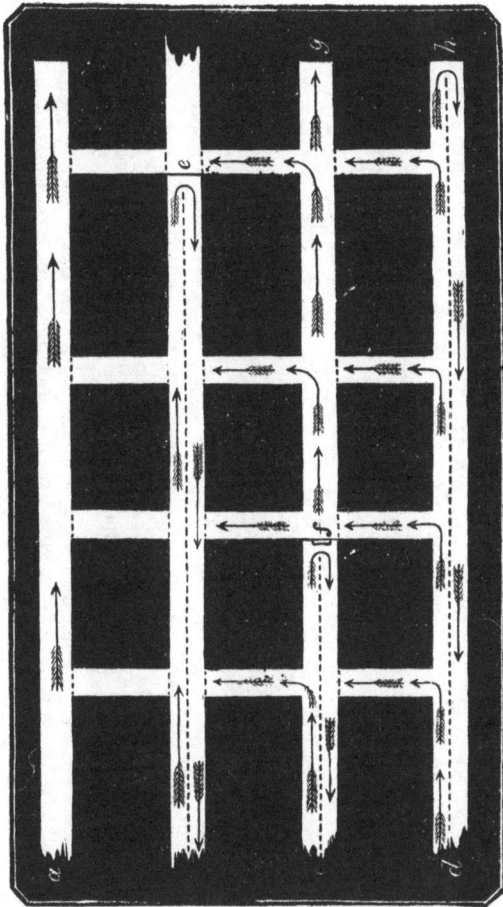

COAL-TOWN STREETS

such places in years that are past, before the way to get good air into these busy streets was understood. An old pitman here and there can remember something of the oppression of head and heart which accompanied his working hours in darkness. Many a "black hole" has held human beings besides that at Calcutta. It was of no use then to complain; good air was not to be had.

COAL-TOWN WORKER.

Well, now for the way of getting air to the black citizens of these black cities. In small pits an open iron grate is filled with coal, kindled, and fixed at the bottom of one of the shafts, which is thereby heated, and rare-faction being produced, you know how the colder air below must rush up the heated vacancy. This method will produce a circulation of a thousand feet of air per minute.

In all larger pits a furnace of large size takes the place of the small grate. A broad, glowing, fiery mass of coals is kept ever burning at the bottom of one shaft, which then takes the place of a vast chimney to the furnace. The other lets in pure air to the mine, while this is the outlet of what has become impure. If you get papa to hold an open newspaper across the front of your fireplace, close to it, you will see how the chimney

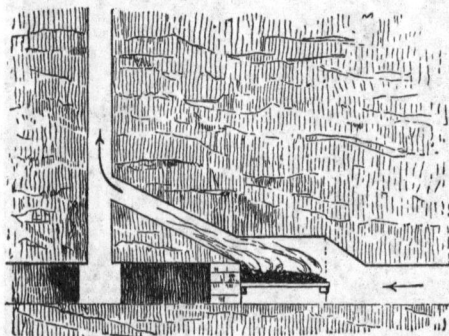

VENTILATING FURNACE.

ventilates your room, in the same way in which this plan ventilates the pit. The chinks in the door and by the windows let in the fresh air, and the heated chimney draws off the impure you have spoilt with breathing.

But how is the air-current to be forced along all streets and alleys, to visit every man in the mine?

The arrangement is this :—

Air is as ductile as water, and for ventilating purposes can be made to take the course man determines. By erecting milldams and stoppings we can divert a stream from its direct course into any other course we may desire, and we can in like manner divert air. By placing barriers of brick, stone, or wood in every desirable position in the coal-town, we can stop the air-current in all such positions, and divert it into others; can compel it

DOOR OF VENTILATING FURNACE.

to advance first up a portion of the main street, and then to turn down a side street to the right hand, or a court to the left. It may turn off at every corner, and go whithersoever man desires; and in this way a single air-current can be taken round the whole pit, be then brought to the upcast shaft, and made to ascend it. The power of the furnace settles that it *must go*, the doors and barriers in its course settle *which way*.

Just so is the coal-pit aired in every part, and the air-currents may be split into as many subdivisions as are necessary, and by this method the whole pit may be benefited.

No less than a hundred thousand cubic feet of air are now passed every minute through the largest northern mine.

The annexed illustration may serve to show how the currents of air circulate in a portion of a coal-mine, in accordance with the arrangements just explained. The dark blocks represent the solid walls of coal which correspond with the houses along streets. All these longitudinal and transverse passages are presumed to have free inter-communication, the only obstacles being the partial stoppages set up for directing or diverting the current of ventilating air. The air is signified by arrows. The wider white lines are the main streets of the pit, called the mainways, and the narrower lines, the side-streets and courts.

In the illustration (p. 191) dotted lines represent such stoppages. In the first long passage, *a*, no stopping is shown in the whole length, and therefore the air rushes straight onward to the end. In the second long passage, *b*, a dotted line runs along to *e*, where there is a full stopping, and where, therefore, the arrows turn round the dotted line, and show the air-current (which had been split by the partition or dotted line into two currents) returning to *b*. In the passage *c*, the same kind of longitudinal division, or dotted line, is carried as far as *f*, and there the split current of air turns round

EXPLOSION IN COAL-MINE.

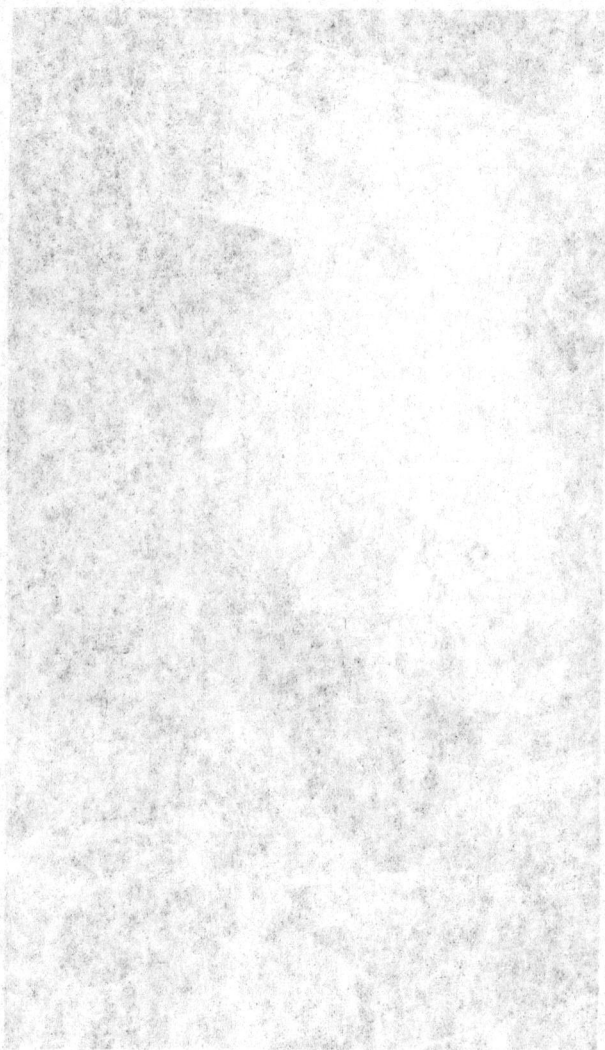

towards *c*. But at *f* there is a door, which, when shut,
drives the air round the dotted line, and when open
permits the air to rush onward to *g*. In the passage *d*,
a longitudinal division is erected for the whole length,
and therefore causes the divided current to start from
d, and when it arrives at *h* to turn round and come back
to *d*.

A large pit, it must be remembered, is but a multiplication of such portions as this plan shows.

And now about the lighting of this subterranean town.
My readers will not think, as an Irishman is said to have
thought, that this will be done by pumping sunlight
down. But the question of lighting is the one great
difficulty and the one great danger of these coal-forest
towns, just because there are what is called " fiery pits."

A " fiery pit," in mining language, is one peculiarly
liable to explosions of the fiery gases.

Of the inflammable gases which infest mines, the most
prevalent and the most dangerous is light carburetted
hydrogen. Look at the thousands of gas-lights illumi-
nating any of our streets at night, and remember that
every one of these is an illustration of the gases that
slumber in coal-seams. A fiery coal-pit, therefore, is
a kind of natural gasometer, where, if there be the
application of a light, a fearful explosion will occur,
blowing up the pit and injuring and killing all persons
near.

Explosions have frequently occurred by the introduction of a lighted candle into cellars and rooms filled with

escaped gas. Some years ago, the windows were blown out, and the walls shattered, in the lower part of a house in Albany Street, Regent's Park. This was merely a house explosion instead of a pit explosion; but explosions in coal-pits are usually far more dreadful than explosions in houses.

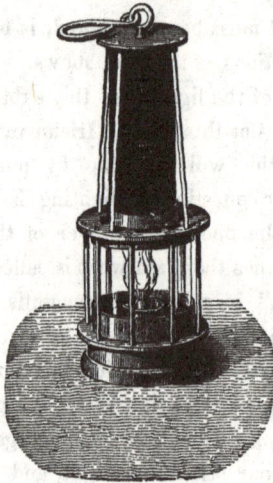

THE DAVY-LAMP.

Some pits, and those commonly the best for household consumption, contain this explosive gas in large quantities. The old Wallsend pit, now closed, was a terribly fiery one.

The most serious cases are those in which the imprisoned gas dislodges a loosened block of coal or stone from its place, and then rushes suddenly and overflow-ingly into the pit, fouling all the passages, and loading the air-currents with its volume. It has been known in this way to spread over a large space in five minutes. In any such condition the introduction of a flame for one second produces an explosion, and all human beings who are within this area will be subjected to burning and death. A few charred lumps, very like a mass of charcoal, will be all that remains of them. Some few may escape singed or scorched in one or more limbs. We have talked with men who have witnessed such scenes, and shuddered to hear their accounts of their narrow escapes from the raging flame. In a subsequent chapter we shall tell one of their stories.

The gaseous coal must and will be wrought. Thousands have wrought it at the expense of their lives, and thou-sands of charred corpses have been committed to graves in the northern churchyards, or even yet sleep unre-covered from the recesses of the exploded pits.

To light the miner and to save his life, Sir Humphry Davy invented his miners' safety-lamp, commonly called the Davy-lamp.

Its operation may be easily understood. It consists of an oil-lamp, enclosed in a cylinder of very fine wire-gauze, of which the apertures should be four hundred and eighty in number to a square inch of surface. In such case the *flame* of the oil-lamp will not pass through the wire gauze, though the *light* will, and thus a moderate illumina-tion and ordinary safety are combined in a portable lamp,

It has been greatly improved by subsequent additions and modifications, and the latest lamps are admirable instruments, reasonably safe, and invaluable companions.

That explosions will occasionally happen is too probable, owing to the enormous and irrepressible eruptions of fire-damp which sometimes burst forth from roof and floor of coal strata; but that explosions may be rather exceptional than evils of regular recurrence in one or other fiery mine, most practical men are ready to admit.

Yet from explosions by the carelessness of the miners and the unavoidable accidents of the pits, it is a melancholy fact that in round numbers annually a thousand English lives are lost.

MINES.

CHAPTER V.

A COAL-MINE—ITS WORK.

WHAT strange sound is that disturbing our slumbers so untimely? Is it a human voice? Is the house on fire, or the pit exploded, or have the French landed on the coast?

No, it is none of these; but the "mine-caller" is going his rounds, or "calling his course," in the village; the labourers are to be roused at dawn, or before it, and as we have engaged to accompany them, it behoves us also to be up betimes. Four o'clock! well, that is early indeed, and it is not yet light. But, what does that matter to men who will descend to darkness? We will not be late, we are up, we are ready, the door is opened, and we stand in the open darkness of the night, seeing nothing but a feeble glimmer or two in cottage windows, hearing nothing but the retreating footsteps and the now distinct voice of the caller.

We made for the pit in company with Dick Fenwick, who now joins us. The shortest way lies across the village and over the fields. As we proceed, our little party is increased by one and another of the hewers who emerge from narrow doors, and pace with hasty step and grave mien to the place of work. How they seem to steal upon us out of the surrounding darkness! Augmented from time to time by one and another in flannel jackets, with bags of *bait* (provisions), and picks and spades upon their shoulders, and frequently pipes in their mouths, our once little company has now become a long file of stragglers, and words of mutual recognition are passing from lip to lip.

Here we are at the pit's mouth, "the bank," as the pitmen call it; and the "cage" is on its way upwards, for us to descend again by it.

Whilst we are waiting for its arrival, we observe how the outward man distinguishes the pitman born and bred, from every other operative you will meet. Diminutive in stature, disproportionate in figure, rather protruding in the chest, having arms rather oddly suspended, and legs sometimes a little bowed, his *tout ensemble* is such as you cannot match. Nor is his countenance less decidedly distinctive than his general figure ; his cheeks are rather hollow, his cheek-bones rather high ; his forehead low, and his brow overhanging. You may occasionally find an agricultural labourer, or blacksmith, or carpenter, or engineer, to whom the familiar term "jolly" might be applied, but I never saw a *jolly* pitman; his work has now made him what you see him, and the high probability is, that if you and I had derived our present existence from mining ancestors, and laboured by candle and lamp light, and in constrained positions, for six hours a day, for

MINER

twenty or thirty years, we should present a similar configuration.

We note, too, that we have ascended an eminence, and now stand upon the "pit-heap," an old and constantly increasing mound of small coal, waste, shale, and general pit-refuse. Those sheds beyond us cover the "screens" through which all the coals pass, and from which comes

all the " Screened Wallsend," so comfortable and profitable in kitchen and parlour. All those wheels and
pulleys above us are connected with the winding machinery
of the mine, and indicate by their revolutions when work

ENGINE HOUSE.

is proceeding, and coals are coming up or men going
down.

Step round this way, and take a look at the old engine-
house. Observe the beam of the pumping-apparatus, half
emerging from an aperture at the top of the house.

Listen for a moment to that long-drawn sound, something between that of a powerful suction of air and a heavy yet measured plunge. That is a stroke of the pumping-gear, —a lifting and lowering of the piston of the steam-engine,

THE CAGE.

—one elevation and outpouring of water from nether "feeders" or springs, that supply five or six hundred gallons of never-failing water every minute, without cessation and without diminution.

Look curiously but cautiously at the men who have

now formed a tolerable throng "on bank." There is
the "banksman," a sort of colliery Chancellor of the
Exchequer, who keeps account of all the incomings and
outgoings of the pit. You can no more escape him than
you can the income-tax gatherer, for he has a lynx eye
and a long arm. Further off are knots of lads and small
boys, who are approaching to be ready for later but yet
longer work than their seniors. We might have a word
or two with them, but for that rapping sound,—a signal
which comes from the clapper just above, and intimates
from the bottom of the shaft that coals are coming up;
for some of the men have already preceded us, and per-
formed work.

At length, up comes the "cage," and you see that it
is an iron framework having two compartments, working on conducting-rods, and forming, in fact, a kind of
verticle railway-carriage. Now the banksman drags out
the coal-laden waggons, and signifies that we may enter
the cage, and take their place. This time it will be a
cage-load of men, and the warning words are shouted
down the shaft, in a drawling voice, "Men a' coming
down."

In we get, and off we go. This is a most unromantic
mode of reaching the depths, for here are we "cabined,
cribbed, confined" in a dirty iron frame. The old
method of descending by the basket was far the more
interesting and exciting. When in the basket you could
oscillate, and run the risk of an upset, or a collision with
another basket, or a bang against the shaft. In this cage

AT BOTTOM OF SHAFT.

you are safe, and yet, perhaps, if iron be constantly deteriorating in these shafts, not too safe. One day a link may be severed, and the cage become a coffin at the bottom of the shaft.

Three or four minutes have sufficed to let us down in the world as much as twelve hundred and fifty feet.

We are now at the bottom of the shaft, and stepping out upon the floor of the mine, we are sadly perplexed by the darkness,—having eyes, but seeing not, and hands, but nothing visible to handle. Groping our way to a prostrate log or stump, we sit still for five minutes, get accustomed to the darkness, find things queer beginning to grow out of it, and feel ourselves less queer than we were five minutes ago.

Now for an underground tour. Let us get up, and go on like men who fear nothing, where fearless men have gone before them.

Equipped for the interior, and preceded by an inferior officer of the mine, who knows its ways and windings as well as you know those of your own street, we may advance courageously, especially along the mainway, which is to the pit what the High Street is to a town. Mark how amply this long passage is excavated, and how well supported by wood-work, brick-work, or stone-work. No railway tunnel is more completely finished, nor is this pit tunnel so unpleasant as those of the railways when filled with smoke from locomotives. The illustration indicates the mainway in perspective.

We soon meet with the *drivers*, or boys who conduct

the laden tubs of coal on long low waggons (rolleys), which are drawn by horses and ponies. The equine race down here have the best of it, for the warm and equable temperature of the mine, the tolerably level run of the roads, and the regular feeds, combine to give them sleek coats, and to insure them a good condition. Horses,

COAL TRAIN

then (and mice), are sure to thrive well in these regions; not, indeed, that the boys who drive the beasts do not also thrive well. Their occupation is not unhealthy; they are in perpetual motion in the best-aired parts of the pit, and they sing snatches of characteristic songs as they last out their short whips, and jump up to or down from their uneasy narrow seats at the head of the coal-tubs. They have neither the lot nor the fate of Phaeton, for they see little of the sun, and their horses never run away; yet they must needs be as cautious as though they were driving the chariot of the sun, since in their falls, though they reach the earth in a second, their maimings and thumps and bruises are neither few nor slight.

But here is a strong wooden door; is there no thoroughfare, and must we go back?

Not at all. Stand here one moment, and listen to the low whistling of the ventilating current of air as it sweeps up to this door and impinges upon it. When this noise is audible and continuous, the ventilation is good; for, as the pitmen say, "the doors are singing."

You knock at the door; it is opened by at present invisible agency, as though we had only to cry, "Open, Sesame!" Pass on, but notice that the moment the door is opened the whistling ceases, because the air-current has gone on unhindered, as we hope to do in a minute. But who opened this door?

Yonder, in the dark corner beyond it, sits the opener and shutter, in the shape of a little boy of from ten to twelve years of age, well known as the "trapper" (opener of a trap-door) in all the collieries. There he sits ten or twelve hours immersed in darkness, and obeying the calls and knocks of all passengers. The waggon-pushers are sometimes rather hard upon him, as he will piteously deplore; but his work is easy, though his con-

finement is long. I have met with trappers of from seven
to nine years of age in several pits ; but, happily, the Act
of Parliament which followed upon the revelations of the
Children's Employment Commission excludes from coal-
mines children under ten, and women altogether, at least
from underground labour. They can and do evade the
Act by assuming male attire ; but I believe the instances
of female subterranean degradation are now very few, and
it is enough that they are illegal.

Who and what are those rough, screeching, shouting,

THE TRAPPER.

and objurgating black lads whom we now encounter in
this dim, rather foul and steaming passage ?

Stand aside, I pray, in this little hollow in the wall, or
you may be crushed, for these rough lads neither see you
nor care for you. Beware as they pass, for Jugger-
nautism under a coal-waggon would indeed be an in-
glorious end. These are *putters* or pushers, the foremost
is dragging and his companions are pushing it. This
labour is terribly severe, but nothing in severity to what
it was when the roads were much inferior, and the

tramways unknown. *Then*, it must have been nearly
the worst and most exhausting labour to which white
people were accustomed. Now it is hard enough, yet not
apparently of such a kind that the wonderfully pliant
human frame cannot adapt itself to it. Such a record,
however, of complaints, and fatigues, and pains of back,
and limbs, and head, and occasional serious injuries, I
never met with, as I obtained from *vivâ voce* examina-
tions of dozen after dozen of these youths.

I should be disposed to believe that there are in this

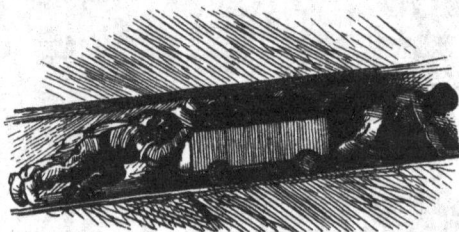

THE PUTTERS.

northern coal-field about eight thousand putters and boys
engaged in the kind of labour we have now witnessed.
I will not revert to the juvenile slavery of former years;
it would only harrow the feelings, and it is now only a
thing of pit history. Yet none but those who have heard
its details, and seen the sort of places in which it was
endured, when the ground of the pit passages was not
laid with metallic plates, can enter into the putter's feelings who gratefully sings—

" God bless the man in peace and plenty
    That first invented metal plates !
Draw out his years te five times twenty,
    Then slide him through the heevenly gates.

" For if the human frame te spare
    Frae toil and pain ayont conceivin !
Hae ought te de wi' gettin' there,
    Aw think he mun gan strite te heeven."

Now for the main body of pit workmen—the original

COAL CUTTING, OLD WAY.

getters of all the coal we have met with in numerous
trains ! These men, I believe, now number in the mines

of the great northern coal-field no less than from thirteen to fourteen thousand, and possibly fifteen thousand. Their labour is of the severest kind, but they work at it only for about six or eight hours " at a spell ; " and it would be impossible to do and endure more. They are amongst

COAL CUTTING, NEW WAY.

the first men in the pit, at three or four in the early morning, and the first out of it. Their workshops are in the innermost recesses of the mine, at the several " faces " or hewing-places of the coal.

Step up to this "face" and watch them at work, as best you may, in an atmosphere heavy with half vitiated air, thick with coal-dust and dark with darkness visible—made visible by two or three consumptive candles, stuck heedlessly above; or, where there is fire-damp, by two or three wire-gauzed, light-giving, Davy-lamps.

See how yonder man squats down, almost frog-like in his attitude, lifts his short pick and drives it straight before him into the seam of coal. He is "nicking out" a line in which to insert a wedge or two, and will then nick

HURRYING.

out others underneath, so that when the wedges are all driven hard in, the coal is fractured, and easily detached by additional blows. The hardness of the work varies as the hardness of the seam, and the strangeness of the posture with the thickness of the seam.

Turn with me into another facing of coal, where the seam is thin, and there you behold the hewer absolutely lying upon his back, supporting his head upon blocks of coal, and using his pick with the utmost constraint, and within the narrowest limits, while he is in constant danger of burying himself in small coal.

We say nothing about clothing in such places and positions, or rather, the lack of it. Our own sensations at this moment prompt us to divest ourselves of all superfluous vestments. But such attitudes as that just represented are exceptional, and become necessary only in working the thin seams.

Still, even under favourable circumstances, hewing is trying and exhausting work; and while you watch the

PICKING.

hewers in their constrained positions, you will conjecture how their odd configuration becomes fixed and hereditary. As you see them bathed in perspiration, you understand why they never look "jolly;" and as you yourself inhale coal-dust, and cough and breathe with difficulty, you will know to what may be traced the colliers' "black-spit" and pulmonary diseases.

I am sure, from what I saw of the principal Newcastle and Durham mines (particularly the newer ones), that much is done to diminish these evils, yet much that is injurious remains unremedied, and probably is not easily remediable. When, however, the mine visitor inspects collieries in certain inferior and much ruder coal-fields, he will no longer be inclined to discredit the testimonies of local medical men, who have frequently testified that the diseases of colliers are distinctly referable to their work and their working places; that an old collier—unless prematurely old—is a rarity; that fifty years of life bring incipient old age, and that added to this, heart diseases and ruptures (caused by over muscular exertion) are not common affections.

Good pay, good feeding, tolerably regular employment, and plenty of free fuel, with other occasional advantages, are some compensation for these evils.

MINES.

CHAPTER VI.

A COAL-MINE ADVENTURE.

SOME years since I paid a visit to a coal district, and one of the entertainments by which my host sought to make my time pass pleasantly was a descent into a coal-mine. I rather liked the idea, as I had never been down one, and at once agreed to go. The mine that was to be visited was an old one of considerable size and depth—the depth of the shaft being nearly one thousand feet. There were some six or eight in our company, and a superior workman—I do not know his proper technical designation—perhaps underground bailiff; at any rate, something equivalent to what we above ground should call the foreman.

I rather repented of my readiness to join the party when I saw the means by which we were to descend, but

I had not courage or time to dissent from what seemed the recognised mode of procedure. We stood in a box and clung to the chain by which we were lowered. No one else seemed to mind it.

" All right ! " said some one, and away we went. My first sensation was that sort of deliquium or swimming in the head that the reader may have experienced when he dreams that he is falling down a precipice. Fortunately it did not relax the muscles, for as it passed away I found myself clinging to the chain like grim death; probably it was only momentary, as I had time to observe the rapidity with which we passed into total darkness.

The story about seeing stars at noonday from the bottom of a coal-pit cannot be true, at any rate if the pit is what is called an up-cast shaft. We went down the up-cast shaft—that is, the shaft by which the air which has entered the pit by the down-cast shaft returns to the upper regions, after having circulated through the mine ; and looking upwards through this air, we could see nothing of the opening of the pit almost immediately after beginning to descend. I suppose the air was so loaded with impurities, coal dust, vitiated vapours, &c., that, seen in quantity, it was as muddy and impenetrable to light as the river Thames at London Bridge, although on the small scale both appear transparent. Down, down, we went, and presently we became aware of a little drizzling rain. It was the water, which, pouring or trickling from the sides of the shaft, sparked off from every projection. As we went deeper this got worse,

DESCENDING THE SHAFT.

and by the time we reached the bottom we were in a
heavy shower.

Suddenly we stopped; we had reached the foot of the
shaft. We found ourselves in the midst of a group of
horses, one of which, a blind old beast, I remember,
came knocking up against me, and nearly upset me.
Some of us were then furnished with lights. I was one
of those that were not. When I say that the lights were
all naked and without protection, the reader will see that
my visit must have been made a good many years ago.

Under the guidance of the foreman we then set off on
our tour. The main passage, along which we went at
first, was what I imagine would be considered a lofty and
spacious gallery, laid with rails. It was comparatively
broad, and seemed to my eye about nine or ten feet high.
We proceeded along this for, I daresay, a quarter of a
mile.

By-and-by our leaders turned into an apparently
unused side gallery, narrower than the main passage, in
which the foreman had something about the ventilation
to point out to the owners. Hitherto we had seen no
men mining; we had met men with horses drawing
trucks, and others going about their occupations, but no
men working. We proceeded along this smaller gallery
for about a hundred and fifty yards or so. The place was
dirty, sloppy, and wet, and, of course, dark; and feeling
no particular interest in what the foreman was desirous of
pointing out to the owners, I lagged behind a little. I
might have been twenty paces behind the rest of the party,

Q

when a sudden light started up among them—I can compare it to nothing but the flash of mimic lightning, with this difference, the light flashed up to the roof and assumed the mushroom shape, but it did not disappear. Instead

MINERS.

of being extinguished as instantaneously as it arose, it continued extending and spreading out along the roof on every side.

My first idea when I saw the light was, that this was some civility on the part of the owners to show off the mysteries of the place to their visitors, as I had seen the Blue-John Mine in Derbyshire, and other stalactitic caves, illuminated by Roman candles and other lights. That idea only lasted for a second.

As the light extended, every one rushed panic-stricken from it as fast as they could run. I guessed the truth in a moment, and turned to fly. There was no difficulty in finding my way, the whole place being illuminated. After flying along for some time I looked back; the whole of the gallery where we had been was one body of fire—not a bright lambent blaze, but lurid, reddish volumes of flame, rolling on like billows of fiery mist. Their form was like that of the volumes of black smoke which we may see at times issuing out of large factory chimneys, more than anything else I can compare it to. My notions of explosions of fire-damp were, that they took place with the rapidity of an explosion of gunpowder. But it was not so in this case, at any rate. I do not mean that it was slow, but that its speed was no greater than that of a man. All those who were at the end of the gallery where it took place did, in point of fact, outrun it. Neither was there any noise or sound of explosion; at least, I noticed none, and if there had been I think I must have observed it, for, all things considered, I was tolerably collected. The report must have taken place at the pit-mouth, as from the mouth of a gun.

The fire rolled silently along in billows of reddish flame,

one wave tumbling over another in quick succession.
And a curious and a very beautiful thing were the edges
of these billows; they were fringed with sparks of blue
flame, dashed off like sparks from a grindstone. Even at
that dreadful moment I could not avoid being struck by
their beauty.

All this I must have gathered at a glance—in an instant
of time. In front of the billowy mass of fire rolling on
towards me I saw the dark figures of my companions
tearing along at headlong speed. Then turning, I again
dashed on. When I came to the loftier main passage I
heard a voice behind me cry out, "Down on your face!"
and by-and-by one figure after another sprang past me
and dashed themselves headlong on the ground. I can
liken the reckless, frantic way in which it was done, to
nothing but boys, when bathing, taking "headers" into
a stream. Without reasoning about it I followed suit,
and flung myself into a puddle, and then peering backwards under my arm, waited the approach of the sea of
flame, the wall of fire, which was approaching. It had
not yet come out of the side gallery, but the glare of its
light preceded it.

Presently it rolled into sight, filling the whole mouth
of the side gallery, from top to bottom. Had it overtaken
us in it, not a soul would have escaped alive; but when
it entered the larger gallery it lifted, just as one sees a
mist lifting on the mountains, and then rolled along the
roof, passing over our heads. How much space there
was between us and it, I cannot say; I imagine it filled

the upper two-thirds, leaving a space of, perhaps, two or three feet free from flame. Nor can I well say how long we lay below this fiery furnace; it might have been five minutes or a quarter of an hour. Judging from our sensations, it must have been hours, but we did not experience so great heat as I should have expected. We felt it more afterwards; probably the anxiety of the moment made us insensible to its intensity.

After the lapse of some time the volume of fire above began to diminish, the stratum got thinner and thinner; it eddied, and curled, and streamed about, leaving the more prominent parts of the roof exposed like islands; then it wandered about like fiery serpents and tongues of flame, licking a corner here, or flickering about a stone there, but ever moving towards the shaft. As it thus abated, presently one head was raised from the ground, then another, until we all began to get up. We then gathered together, but there were no mutual congratulations, nor external acknowledgment of thanks to God, however much some may have felt. But I doubt if there was much feeling of that kind, the sense of peril was yet too strong; we had escaped one great danger, but we knew that we were still exposed to the risk of many others which often followed such explosions. The first danger was want of air; the fire had used what was in the mine almost wholly up, and we might perish from want of it.

"Follow me!" said the foreman, and he started off, not for the mouth of the mine, but for some part of it which, from its connections or position, he knew to be better, or more likely to be supplied with air than any other part. The miners knew this too, doubtless, for on our arrival at the place in question, we found them trooping in from different quarters, until there might be above a hundred present; and I was much struck by one thing in them which was not according to my anticipations. I thought that men who were habitually exposed to any danger became callous to it, and faced it with indifference. It was not so with these miners; we, who scarcely understood the magnitude of the danger through which we had passed, were far cooler and more collected than they. Almost every one of them was thoroughly unmanned, and shook in every fibre. I know the ague well, and the uncontrollable shaking which bids defiance to the strongest exercise of the will, but I never saw a worse tremor in ague than in these men.

While gathered together in this part of the mine a loud crack ran through the roof above our heads, which so alarmed the already nerveless miners that some of them actually sunk upon the ground. The explanation of this anomaly in men's courage is, I think, that where they see their danger, and can exert themselves to ward it off or escape it, familiarity with it will produce contempt for it; but where they are utterly helpless, and know that they are so, familiarity with it only adds to its terrors. This is the case with earthquakes. No familiarity with them enables a man to meet them with composure; the more he has felt, the more frightened he becomes.

After waiting a considerable time in this part of the mine—perhaps an hour—we again started, and made for the mouth of the pit. As we approached it we heard shouts, and presently came upon a body of men, who, having heard the explosion, had been sent down to see what mischief had been done. Although the explosion had travelled so deliberately when it passed over us, it had had sufficient violence when it reached the shaft to blow the roof of the building adjoining the pit-mouth clean off. Fortunately, it had not destroyed the gear there, and we were able to ascend without delay. Right glad was I to find myself once more in the open air. The explosion had drawn a crowd of agitated men and women to the mouth of the mine.

Alas! the meaning of the dull report, and the cloud of smoke, and the fragments of the building at the pit's mouth flying in the air, were too well known in the neighbourhood, and many an anxious heart found relief in a burst of tears when we were able to announce, on our appearance at the surface, that no lives had been lost. We escaped with almost miraculously slight injury for men who had gone through an explosion of fire-damp.

MINES.

CHAPTER VII.

ANOTHER COAL-MINE ADVENTURE.

AN incident, perhaps one of the most tragic in the history of coal-mining in this country, occurred in the little Girvan coal-field.

In the quiet churchyard of Dailly lie the unmarked graves of generations of colliers; but among them is one with a tombstone bearing the following inscription: "In memory of John Brown, Collier, who was enclosed in Kilgrammie coal-pit, by a portion of it having fallen in, October 8th, 1835, and was taken out alive, and in full possession of his mental faculties, but in a very exhausted state, October 31st, having been twenty-three days in utter seclusion from the world, and without a particle of food. He lived for three days after, having

quietly expired on the evening of November 3rd, aged
sixty-six years."

Three weeks without food in the depths of the earth!

The following narrative is compiled from an account
given by Professor Geikie.

For two days cracks and grinding noises had been
heard continuously in the pit, the levels of the rails got
more and more altered, and though the men remained at
work, it became hourly more clear that part of the workings would now need to be abandoned.

At last, on the 8th of October, the final crash came
suddenly and violently. The huge weight of rock under
which the galleries ran settled down solidly on them with
a noise and shock which, spreading for a mile or two up
and down the quiet vale of the Girvan, were set down at
the time as the passing of an earthquake. Over the site
of the mine itself the ground was split open into huge
rents for a space of several acres, the dam of a pond gave
way, and the water rushed off, while the horses at the
mouth of the pit took fright, and came scampering,
masterless and in terror, into the little village, the inhabitants of which rushed out of doors, and were standing in
wonderment as to what had happened.

But the disasters above ground were only a feeble indi-
cation of the terrors underneath.

Constant exposure to risk hardens a man against an
appreciation of his dangers, and even makes him, it may
be, foolhardy. The Kilgrammie colliers had continued
their work with reckless disregard of consequences, until

at last the cry arose among them that the roof was settling down. First they made a rush to the bottom of the shaft, in hopes of being pulled up by the engine. But by this time the shaft had become involved in the ruin of the roof. A second shaft stood at a little distance ; but this too they found to be closed. Every avenue of escape cut off, and amid the hideous groanings and grindings of the sunken ground, the colliers had retreated to a part of the workings where the pillars stood firm.

Fortunately, one of them remembered an old tunnel, or " day-level," running from the mine for more than half a mile to the Brunston Holm, on the banks of the Girvan, and made originally to carry off the underground water. They were starting to find the entrance to this tunnel, when they noticed, for the first time, that John Brown was not among them. Two of the younger men started back through the falling part of the workings, and found the old man at his post, working as unconcernedly as if he had been digging potatoes in his own garden. With some difficulty they persuaded him to return with them, and were in the act of hurrying him along, when he remembered that in the haste he had left his jacket behind. In vain they tried to drag him along. "The jacket was a new one," he said ; "and as for the pit, he had been at a crush before now, and would win through it this time too." So, with a spring backwards he tore himself away from them, and dived into the darkness of the mine in search of his valued garment. Hardly, however, had he parted from them, when the roof between him and them

came down with a crash.   They managed to rejoin their
comrades ; John Brown was sealed up within the mine,
most probably, as they thought, crushed to death between
the ruins of the roof and floor.

Those who have ever by any chance peeped into the
sombre mouth of the day-level of a coal-pit, will realise
what the colliers had now to do to make good their
escape.   The tunnel had been cut simply as a drain ;
dark water and mud filled it almost to the roof.

For more than half a mile they had to walk, or rather
to crouch along in a stooping posture through this con-
duit, the water often up to their shoulders ; sometimes,
indeed, with barely room for their heads to pass between
the surface of the slimy water and the roof above.   But
at length they reached the bright daylight as it streamed
over the green holms and autumn woods of the Girvan, no
man missing save him whom they had done their best to
rescue.   They were the first to bring the tidings of their
escape to the terrified village.

No attempt could at first be made to save the poor
fellow.   As the colliers themselves said, not even a creel,
or little coal-basket, could get down the crushed shaft of
the pit.

After the lapse of some days the cracking and groaning
of the broken roof had so far abated, that it became possible once more to get down into the pit.   The first
efforts were, of course, directed towards that part of the
workings where the body was believed to be lying.   But
the former roadways were found to be so completely

blocked up, that no approach to the place could be had
save by cutting a new tunnel through the ruins. This
proved to be a work of great labour and difficulty ; for
not only were the materials extremely hard through which
the new passage must be cut : a dead body lay in the pit,
and awakened all the superstition of the colliers. At
times they would work well, but their ears were ever on
the alert for strange, weird noises, and often would they
come rushing out from the working in terror at the
unearthly gibberings which ever and anon would go
soughing through the mine.

A fortnight had passed away.

The lessee, like the rest of the inhabitants, believed
poor Brown to be already dead, and brought a gang of
colliers from another part of the country to help in
clearing out and re-opening his coal-pit. But a party of
the men continued at work upon the tunnel that was to
lead to the body. They cut through the hard crushed
roof a long passage, just wide enough to let a man crawl
along it upon his elbows : and at last, early on the
morning of the twenty-third day of the accident, they
struck through the last part of the ruined mass into the
open workings beyond.

The rush of foul air from these workings put out their
lights, and compelled them to retreat. One of their
number was dispatched to upper air for a couple of
boards, or corn-sieves, or any broad flat things he could
lay hands upon, with which they might advance into the
workings, and waft the air about, so as to mix it, and

AFTER A CRUSH.

make it more breathable. Some time had to elapse before the messenger could make the circuitous journey, and meanwhile the foulness of the air had probably lessened.

When the sieves came, one of the miners agreed to advance into the darkness, and try to create a current of air ; the rest were to follow. In winnowing the air with his arms, he had struck against a waggon standing on the roadway, and the noise he had made was followed by a distinct groan. A younger member of the gang volunteered to return with him. Advancing as before, the same waggon stopped them as their sieves came against the end of it, and again there rose from out of the darkness of the mine a faint, but audible, groan. Could it be the poor castaway, or was it only another wile of the arch-enemy to lure two colliers more to their fate ?

Gathering up all the courage that was left in him, one of them broke the awful silence of the place by solemnly demanding, "If that's your ain groan, John Brown, in the name o' God, gie anither." They listened, and after the echoes of his voice had ceased they heard another groan, coming apparently from the roadway only a few yards ahead. They crept forward, and found their companion—alive.

In a few seconds the other colliers, who had been anxiously awaiting the result, were also beside the body of John Brown. They could not see it, for they had not yet resumed their lights ; but they could feel that it had the death-like chill of a corpse. Stripping off their

jackets and shirts, they lay with their naked backs next
to his, trying to restore a little warmth to his hardly
living frame. His first words, uttered in a scarcely
audible whisper, were, "Gie me a drink." Fearful of
endangering the life which they had been the means of so
marvellously saving, they only complied so far with his
wish as to dip the sleeve of a coat in one of the little
runnels which were trickling down the walls of the mine,
and to moisten his lips with it. He pushed it from him,
asking them "no to mak' a fule o' him." A little water
refreshed him, and then in the same strangely sepulchral
whisper, he said, "Eh, boys, but ye've been lang o'
coming."

Word was now sent to the outer world that John
Brown had been found, and was yet living. The lessee
came down, the doctor was sent for, and preparations
were made to have the sufferer taken up to daylight
again. And here one of the strangest parts of the story
must be told.

If by chance the reader has ever been in a coal-pit, he
may have remarked that upon the decayed timber props
and old wooden boardings an unseemly growth of a
white and yellow fungus often takes root, hanging in
loathsome tufts and bunches from the sides or roofs
wherever the wood is decaying. After being cautiously
pushed through the newly-cut passage, John Brown was
placed on the lessee's knees on the cage in which they
were to be pulled up by the engine. As they rose into
daylight, a sight which had only been faintly visible in

the feeble lamp-light below presented itself, never seen
before, and never to be forgotten.   That same loathsome fungus had spread over the poor collier's body as
it would have done over a rotting log.   His beard had
grown bristly during his confinement, and all through the
hairs this white fungus had taken root.   His master, as
the approaching daylight made the growth more visible,
began to pull off the fungus threads, but his hand was
pushed aside by John, who asked him, "Na, noo, wad
ye kittle (tickle) me?"

By nine o'clock on that Friday morning, three-and-
twenty days after he had walked out of his cottage for
the last time, John Brown was once more resting on
his own bed.   A more ghastly figure could hardly be
pictured.   His face had not the pallor of a fainting fit
or of death, but wore a strange sallow hue like that of a
mummy.   His flesh seemed entirely gone, nothing left
but the bones, under a thin covering of leather-like skin.
This was specially marked about his face, where, in spite
of the growth of hair, every bone looked as if it were
coming through the skin, and his eyes, brightened into
unnatural lustre, were sunk far into his skull.   To such a
degree was the body wasted, that in putting the hand
over the pit of the stomach, one could distinctly feel the
inner surface of the backbone.   Every atom of fatty
matter in the body seemed to have been consumed.

On the Sunday afternoon, when some of his old comrades were sitting round the bedside, he turned to them
with an anxious, puzzled look and said, "Ah, boys, when

R

I win through this, I've a queer story to tell ye." But
that was not to be. His constitution had received such a
shake as even its uncommon strength could not overcome.
That evening it became only too plain that the apparent
recovery of appetite and spirits had been but the last
flicker of the lamp of life. Later in the night he died.

QUARRIES

QUARRIES.

CHAPTER I.

SANDSTONE, ITS INHABITANTS.

LET us take a peep into a sandstone quarry. The so-called New Red sandstone, which, with its attendant marls, covers a vast tract—and that a rich and busy one —of England. From Hartlepool and the mouth of the Tees, down through Yorkshire and Nottinghamshire; over the manufacturing districts of central England; down the valley of the Severn; past Bristol and the Somersetshire flats to Torquay in South Devon; up north-westward through Shropshire and Cheshire; past Liverpool and northward through Lancashire; reappearing again, north of the Lake Mountains, about Carlisle and the Scotch side of the Solway Frith—stretches the great New Red sand-stone plain, from under which everywhere the coal-bearing rocks rise as from a sea.

It contains, in many places, excellent quarries of building stone; the most famous of which, perhaps, are the well-known Runcorn quarries, near Liverpool, from which the old Romans brought the material for the walls and temples of ancient Chester.

It is five thousand feet thick in places, and stretches not only across England, but also into Germany.

Whence came this enormous mass of sand?

It is difficult to answer. The shape and distribution of land in the days when the New Red sandstones were being formed, were so different from what they are now, that the rocks which furnished a great deal of our sandstone may be now, for aught I know, a mile beneath the sea.

But over the land which still stands out of the sea near us, there has been wear and tear enough to account for any quantity of sand deposit.

As a single instance: It is a provable and proven fact —as you may see from Mr. Ramsay's survey of North Wales—that over a large tract to the south of Snowdon, between Port Madoc and Barmouth, there has been ground off and carried away a mass of solid rock twenty thousand feet thick; thick enough, in fact, if it were there still, to make a range of mountains as high as the Andes.

Again, it is a provable and proven fact that vast tracts of the centre of poor old Ireland were once covered with coal-measures, which have been scraped off in like wise, deprived of inestimable mineral wealth. The destruction of rocks—" denudation " as it is called—in the district round Malvern, is, I am told, probably enormous.

Indeed, it is so over all Wales, North England, and West
and North Scotland. So there is enough of rubbish to be
accounted for to make our New Red sands.

For you find in the beds, from the top to the bottom
(at least in Cheshire), particles of mica. Now this mica
could not have been formed in the sand. It is a definite
crystalline mineral, whose composition is well known.
It is only found in rocks which have been subjected to

QUARRY SHOWING VARIOUS BEDS.

immense pressure, and probably to heat. The granites
and mica-slates of Anglesey are full of it; and from
Anglesey—as likely as from anywhere else—these thin
scales of mica came. It is certain, too, that most of these
sands were deposited in a very shallow water, and very
near to land, for sand and pebbles could not be carried far
out to sea; it is further certain that their surface was

often out of water. Of that you may see very pretty proofs.

You find these sands ripple-marked, as you do shore-sands now. You find cracks where the marl mud has dried in the sun; and, more, you find the little pits made by rain. Of that I have no doubt. I have seen specimens in which you could not only see at a glance that the marks had been made by the large drops of a shower, but see also from what direction the showers had come.

And then the delicate markings were covered up with a fresh layer of mud or sand. How long since? How long since that flag had seen the light of the sun, when it saw it once again, restored to the upper air by the pick of the quarryman? Who can answer that? Not I.

Fossils are very rare in these sands; it is not easy to say why. But what will interest the reader are the footprints of a strange beast, found alike in England and in Germany—the Cheirotherium, as it was first named, from its hand-like feet; the Labyrinthodon, as it is now named, from the extraordinary structure of its teeth. There is little doubt now, among anatomists, that the bones and teeth of the so-called Labyrinthodon belong to the animal which made the footprints. If so, the creature must have been a right loathy monster. Some think him to have been akin to lizards; but the usual opinion is that he was a cousin of frogs and toads.

Looking at his hands and other remains, one pictures him to oneself as a short, squat brute, as big as a fat hog, with a somewhat hog-like head, and very large, baboon-like

hands behind and small ones in front, waddling about on
the tide flats of a sandy sea, and dragging after him,

THE LABYRINTHODON RESTORED.

seemingly, a short tail, which has left its mark on the
sand. What his colour was, whether he was smooth or

warty, what he ate, and in general how he got his living, we know not. But there must have been something there for him to eat; and I dare say that he was about as happy and about as intellectual as a toad is now.

Remember always that there is nothing alive now exactly like him, or, indeed, like any animal found in these sandstones. The whole animal world of this planet

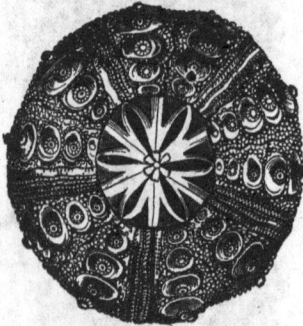

URCHIN, FOSSIL.

has changed entirely more than once since the Labyrinthodon waddled over the Cheshire flats.

But there is a more remarkable animal of which I must say a few words, and one which to scientific men is most interesting and significant.

Both near Warwick, and near Elgin in Scotland, in Central India, and in South Africa, fossil remains are found of a family of lizards, utterly unlike anything now living save one, and that one is crawling about plentifully, I believe—of all places in the world—in New Zealand.

How it got there ; how so strange a type of creature should
have died out over the rest of the world, and yet have
lasted on in that remote island for long ages, ever since
the days of the New Red sandstone, is one of those questions—quite awful questions I consider them—with which
I will not puzzle my readers.   I only mention it to show

AMMONITE, FOSSIL.

them what serious questions the scientific man has to face
and to answer, if he can.

Only the next time they go to the Zoological Gardens
in London, let them go to the reptile-house, and ask the
very clever and courteous attendant to show them the
Sphenodons, or Hatterias, as he will probably call them
—and then look, I hope with kindly interest, at the oldest
Conservatives they ever saw, or are likely to see ; gentle-
men of most ancient pedigree, who have remained all but

unchanged, while the whole surface of the globe has changed around them more than once or twice.

And now, of course, my readers will expect to hear

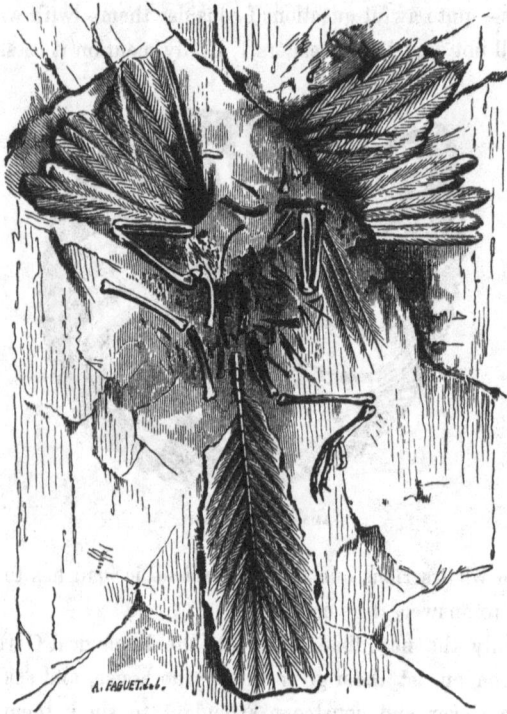

FOSSIL BIRD.

something of the deposits of rock-salt, for which Cheshire and its red rocks are famous. I have never seen them, and can only say that the salt does not, it is said by geologists, lie in the sandstone, but at the bottom of the red marl which caps the sandstone.   How was it formed ?

Most probably by the gradual drying up of lagoons, such as are depositing salt, it is said now, both in the Gulf of Tadjara, on the Abyssinian frontier opposite Aden, and in the Runn of Cutch, near the Delta of the Indus. If this be so, then these New Red sandstones may be the remains of a whole Sahara—a sheet of sandy and all but lifeless deserts, reaching from the west of England into Germany, and rising slowly out of the sea; to sink, as we shall find, beneath the sea again.

And now, as to the vast period of time—the four or five worlds, as I called it—which elapsed between the laying down of the New Red sandstones and the laying down of the Boulder clays.

I think this fact—for fact it is—may be better proved by taking the readers an imaginary railway journey to London from any spot in the manufacturing districts of central England—begging them, meanwhile, to keep their eyes open on the way.

And here I must say that I wish folks in general would keep their eyes a little more open when they travel by rail.

When I see young people rolling along in a luxurious carriage, their eyes and their brains absorbed probably in a trashy shilling novel, and never lifted up to look out of the window, unconscious of all that they are passing—of the reverend antiquities, the admirable agriculture, the rich and peaceful scenery, the like of which no country upon earth can show; unconscious, too, of how much they might learn of botany and zoology, by simply watching the flowers along the railway banks and the sections in the

PLESIOSAURUS, FOSSIL.

cuttings : then it grieves me to see what little use people make of the eyes and of the understanding which God has given them. They complain of a dull journey; but it is not the journey which is dull: it is they who are dull. Eyes have they, and see not; ears have they, and hear

not; mere dolls in smart clothes, too many of them, like
the idols of the heathen.

But my readers, I trust, are of a better mind.   So the
next time they find themselves running up southward to
London—or the reverse way—let them keep their eyes

PTERODACTYLE, FOSSIL.

open, and verify, with the help of a geological map, the
sketch which is given in the following page.

The "Black Countries"—the actual coal districts—are
in England either shores or islands yet undestroyed, which
stand out of the great sea of New Red sandstone.   But
with them I will not confuse the reader just now, but will

only ask him to keep his eye on the rolling plain of New
Red sands and marls past, say, Birmingham and Warwick.
After those places, these sands and marls dip to the south-
east, and other rocks and soils appear above them, one
after another, dipping likewise toward the south-east—
that is, toward London.

First appear thin layers of a very hard blue limestone,
full of shells, and parted by layers of blue mud.   That
rock runs in a broad belt across England, from Whitby in
Yorkshire, to Lyme in Dorsetshire, and is known as Lias.
Famous it is, as some readers may know, for holding the
bones of extinct monsters—Ichthyosaurs (Fish-Lizard)
and Plesiosaurs, resembling a lizard such as the unlearned
may behold in the lake at the Crystal Palace. On this
rock lie the rich cheese pastures and the best tracts of
the famous " hunting shires " of England.

Lying on it, as we go south-eastward, appear alternate
beds of sandy limestone, with vast depths of clay between
them.   These " Oolites," or freestones, furnish the famous
Bath stone, abound in fossils, but of kinds, be it remembered, which differ more and more from those of the Lias
beneath.   There, too, are found principally the bones of
that extraordinary flying lizard, the Pterodactyle, (winged-
fingers), which had wings formed out of its forelegs, somewhat like those of a bat ; but with one exception. In
the bat, as any one may see, four fingers of the hand are
lengthened to carry the wing, while the first alone is left
free, as a thumb ; but in the Pterodactyle, the outer or
" little " finger alone is lengthened, and the other four

fingers left free—one of those strange instances in nature of the same effect being produced in widely different plants and animals, and yet by slightly different means, on which

LANDSCAPE OF THE SANDSTONE AGE.

a whole chapter of natural philosophy—say, rather, natural theology—will have to be written some day.

s

QUARRIES.

CHAPTER II.

SANDSTONE, ITS AGE.

BUT now consider what this Lias, and the Oolites and clays upon it, mean. They mean that the New Red sandstone, after it had been dry land, or all but dry land (as is proved by the footprints of animals and the deposits of salt), was sunk again beneath the sea.

Each deposit of limestone signifies a long period of time, during which that sea was pure enough to allow reefs of coral to grow, and shells to propagate, at the bottom.

Each great band of clay signifies a long period, during which fine mud was brought down from some wasting land in the neighbourhood. And that land was not far distant is proved by the bones of the Pterodactyle, of Crocodiles, and of Marsupials; by the fact that the shells are of shallow water or shore species; by the presence,

PTERODACTYLE, RESTORED.

mixed with them, of fragments of wood, impressions of plants, and even wing shells of beetles; and, lastly, if further proof was needed, by the fact that in the "dirt-bed" of the Isle of Portland and the neighbouring shore the stumps of trees allied to the modern sago-palms are found as they grew in the soil, which, with them, has been covered up in layers of freshwater slate and limestone. A tropic forest has plainly sunk beneath a lagoon; and that lagoon, again, beneath the sea.

And how long did this period of slow sinking go on?

Who can tell? The thickness of the Lias and Oolites together cannot be less than a thousand feet. Considering, then, the length of time required to lay down a thousand feet of strata, and considering the vast difference between the animals found in them, and the few found in the New Red sandstone, we have a right to call them another world, and that one which must have lasted for ages.

After we pass Oxford, or the Vale of Aylesbury, we enter yet another world.

We come to a bed of sand, under which the freestones and their adjoining clays dip to the south-east. This is called commonly the Lower Greensand, though it is not green, but rich iron-red.

Then succeeds a band of stiff blue clay, called the gault, and then another bed of sand, the Upper Greensand, which is more worthy of the name, for it does carry, in most places, a band of green or "glauconite"

sand. But this and the upper layers of the Lower
Greensand also, are worth our attention; for we are
all probably eating them from time to time in the form
of bran.

It had been long remarked that certain parts of these
beds carried admirable wheat-land; it had been re-

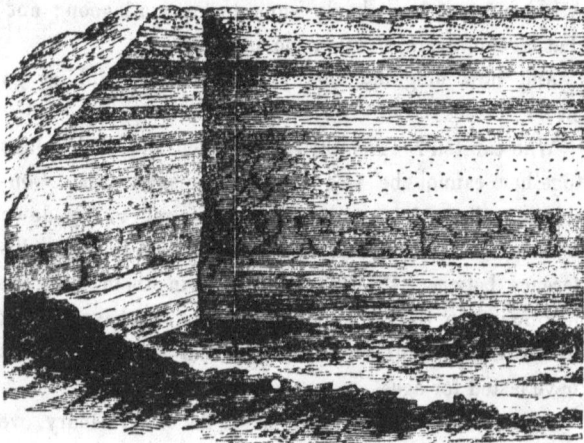

"BEDS."—SAMPLE OF THE EARTH'S CRUST.

marked, too, that the finest hop-lands—those of Farnham,
for instance, and Tunbridge—lay upon them, but that
the fertile band was very narrow—that, as in the Surrey
moors, vast sheets of the Lower Greensand were not
worth cultivation. What caused the striking difference?

Dr. Henslow had brought to him by a farmer (so the

story ran) a few fossils. He saw, being somewhat of a geologist and chemist, that they were not, as fossils usually are, carbonate of lime, but phosphate of lime—that is, bone-earth. He said at once, as by an inspiration, "You have found a treasure—not a gold-mine, indeed, but a food-mine. Only find enough of them, and you will increase immensely the food-supply of England,

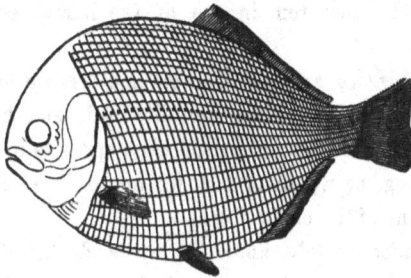

FISH, FOSSIL.

and perhaps make her independent of foreign phosphates in case of war."

His advice was acted on; for the British farmer is by no means the stupid personage which townsfolk are too apt to fancy him. This bed of phosphates was found everywhere in the Greensand, underlying the Chalk. These so-called "coprolites" are used perpetually now for manure. They are fossil excrements of fishes and saurians. The name is from *kopros*, "dung," and *lithos*, "a stone."

So much for the useless "hobby," as some fancy it.

of poking over old bones and stones, and learning a little
of the composition of this earth on which God has
placed us.

How to explain the presence of this vast mass of
animal matter, in one or two thin bands right across
England, I know not. That the fossils have been rolled
on a sea-beach is plain to those who look at them. But
what caused so vast a destruction of animal life along
that beach must remain one of the buried secrets of
the past.

And now we are fast nearing another world, which
is far younger than that coprolite bed, and has been
formed under circumstances the most opposite to it. We
are nearing, by whatever rail we approach London, the
escarpment of the chalk downs.

All readers, surely, know the white chalk, the special
feature and the special pride of the south of England.
All know its softly-rounded downs, its vast beech woods,
its short and sweet turf, its snowy cliffs, which have
given—so some say—to the whole island the name of
Albion—the white land. But all do not, perhaps, know
that, till we get to the chalk, no single plant or animal
has been found which is exactly like any plant or animal
now known to be living. The plants and animals grow,
on the whole, more and more like our living forms as we
rise in the series of beds. But only above the chalk (as
far as we yet know) do we begin to find species identical
with those living now.

This in itself would prove a vast lapse of time. We

shall have a further proof of that vast lapse if we
examine the chalk itself. It is composed—of this there
is now no doubt—almost entirely of the shells of minute
animalcules; and animalcules (I use an unscientific word
for the sake of unscientific readers) like these, and in
some cases identical with them, are now forming a
similar deposit of mud, at vast depths, over the greater
part of the Atlantic sea-floor. This fact has been put
out of doubt by recent deep-sea dredgings. A whole
literature has been written on it of late. Any reader
who wishes to know it, need only ask the first geologist
he meets; and if he has the wholesome instinct of
wonder in him, fill his imagination with true wonders,
more grand and strange than he is like to find in any
fairy-tale. All I have to do with the matter here, is
to say that, arguing from the known to the unknown,
from the Atlantic deep-sea ooze which we do know
about, to the chalk which we do not know about, the
whole of the chalk must have been laid down at the
bottom of a deep and still ocean, far out of the reach
of winds, tides, and even currents, as a great part of the
Atlantic sea-floor is at this day.

Prodigious! says the reader.

And so it is. Prodigious to think that that shallow
Greensand shore, strewed with dead animals, should
sink to the bottom of an ocean, perhaps a mile, perhaps
some four miles deep. Prodigious the time during which
it must have lain as a still ocean-floor. For so minute
are the living atomies which form the ooze, that an

inch, I should say, is as much as we can allow for their yearly deposit; and the chalk is at least a thousand feet thick. It may have taken, therefore, twelve thousand years to form the chalk alone. A rough guess, of course, but one as likely to be two or three times too little, as

MINUTE SHELLS IN CHALK, GREATLY MAGNIFIED

two or three times too big. But such, or somewhat such, is the fact.

Thus, surely, if we call the Oolitic beds one new world above the New Red sandstone, we must call the chalk a second new world in like wise.

I will not trouble the reader here with the reasons
why geologists connect the chalk with the greensands
below it, by regular gradations, in spite of the enormous
downward leap, from sea-shore to deep ocean, which
the beds seem (but only seem) to have taken. The

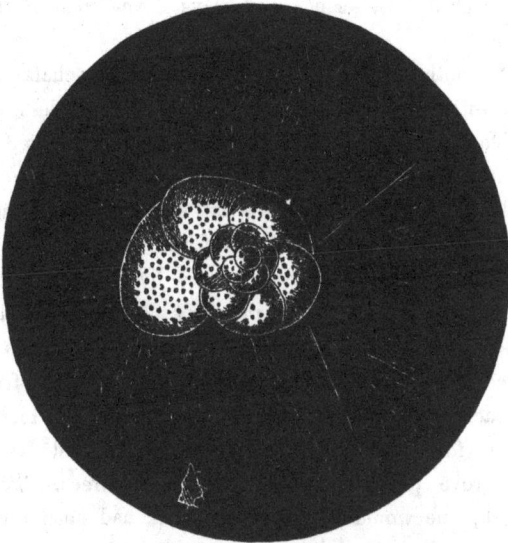

A CHALK-SHELL ANIMAL, RESTORED.

change—like all changes in geology—was probably
gradual. Not by spasmodic leaps and starts, but slowly
and stately, as befits a God of order, of patience, and of
strength, have these great deeds been done.

But we have not yet done with new worlds or new

prodigies on our way to London, as any Londoner
may ascertain for himself, if he will run out a few miles
by rail, and look in any cutting or pit, where the surface of the chalk, and the beds which lie on it, are
exposed.

On the chalk lie—especially in the Blackheath and
Woolwich district—sands and clays. And what do they
tell us ?

Of another new world, in which the chalk has
been lifted up again, to form gradually, doubtless, and
at different points in succession, the shore of a sea.

But what proof is there of this ?

The surface of the chalk is not flat and smooth, as it
must have been at the bottom of the sea. It is eaten
out into holes and furrows, plainly by the gnawing of
the waves; and on it lie, in many places, large rolled
flints out of chalk that has been destroyed, beds of
shore-shingle, beds of oysters, lying as they grew, fresh
or brackish water-shells standing as they lived, bits of
lignite (fossil wood half turned to coal), and (as in
Katesgrove pits at Reading) leaves of trees. Proof
enough, one would say, that the chalk had been raised
till part of it at least became dry land, and carried vegetation.

And yet we have not done. There is another world
to tell of yet.

For these beds (known as the Woolwich and Reading
beds) dip under that vast bed of London clay, four
hundred and more feet thick, which was certainly laid

down by the estuary of some great tropic river, among palm trees and Anonas, crocodiles and turtles.

Is the reader's power of belief exhausted?

If not, there are to be seen, capping almost every

RIVER, ANCIENT ENGLAND.

high land round London, the remains of a fifth world. Some of my readers may have been to Ascot races, or to Aldershot camp, and may recollect the table-land of the sandy moors, perfectly flat atop, dreary enough to those to whom they are not (as they have long been to me) a home and a work-field. Those sands are several hundred feet thick. They lie on the London clay.

And they represent—the reader must take geologists' word for it—a series of, in some places, thousands of feet thick, in the Isle of Wight, in the Paris basin, in the volcanic country of the Auvergne, in Switzerland, in Italy; a period during which the land must at first have swarmed with forms of tropic life, and then grown—but very gradually—more temperate, and then colder and colder still; till at last set in that age of ice, which spread the boulder pebbles over all rocks and soils indiscriminately, from the Lake Mountains to within a few miles of London.

For everywhere about those Ascot moors, the top of the sands has been ploughed by shore-ice in winter, as they lay a-wash in the shallow sea; and over them, in many places, is spread a thin sheet of ice gravel, more ancient, the best geologists think, than the boulder and the boulder-clay.

And if any reader asks—how long the period was during which those sands of Ascot Heath and Aldershot have been laid down, I cannot tell. But this we can tell. It was long enough to see such changes in land and sea, that maps representing Europe during the greater part of that period (as far as we can guess at it) look no more like Europe than like America or the South Sea Islands. And this we can tell besides: that

that period was long enough for the Swiss Alps to be lifted up at least ten thousand feet of their present height. And that was a work which—though God could, if He willed it, have done it in a single day—we have proof positive was not done in less than ages, besides which the mortal life of man is as the life of the gnat which dances in the sun.

And all this and more happened between the date of the boulder-clay, and that of the New Red sandstone on which it rests.

QUARRIES.

CHAPTER III.

SLATE.

SLATES come from the hills of Cumberland, Westmore-land, and North Wales. How beautiful and how curious an object is a common slate!

Beautiful: not only for the compactness and delicacy of its texture, and for the regularity and smoothness of its surface, but still more for its colour.

Beautiful is the slate, and curious likewise, venerable; a most ancient and elaborate work of God. And yet it was at first nought but an ugly lump of soft and shapeless ooze, for slate is hardened mud.

That slate may have been once mud, is made probable by the simple fact that it can be turned into mud again. If you grind up slate, and then analyze it, you will find its mineral constituents to be exactly those of a fine, rich, and tenacious clay.

But, granting this, how did the first change take place ?

It must be allowed, at starting, that time enough has elapsed, and events enough have happened, since our supposed mud began first to become slate, to allow of many and strange transformations. For these slates are found in the oldest beds of rocks, save one series, in the known world; and it is notorious that the older and lower the beds in which the slates are found, the better, that is, the more perfectly elaborate, is the slate.

Any one who saw that noble peak of Snowdon, which is a slate-bearing mountain, leaping high into the air, dominating all the country round, at least upon three sides, would have the right at first sight, on hearing of earthquake faults and upheavals, to say—the peak of Snowdon has been upheaved to its present height above and out of the lower lands around. But when he came to examine sections, he would find his reasonable guess utterly wrong.

Snowdon is no swelling up of the earth's crust. The beds do not, as they would, in that case, slope up to it. They slope up from it, to the north-west in one direction, and the south-south-west in the other ; and Snowdon is a mere insignificant boss, left hanging on one slope of what was once an enormous trough, or valley, of strata far older than itself. By restoring these strata, in the direction of the angles, in which they crop out, and

T

vanish at the surface, it is found that to the north-west—
the direction of the Menai Straits—they must once have
risen to a height of at least six or seven thousand feet;
and more, by restoring them, specially the ash-bed of
Snowdon, towards the south-east—which can be done by
the guidance of certain patches of it left on other hills—it
is found that south of Ffestiniog, where the Cambrian
rocks rise again to the surface, the south side of the trough
must have sloped upwards to a height of from fifteen to
twenty thousand feet, whether at the bottom of the sea, or
in the upper air, we cannot tell. But the fact is certain,
that off the surface of Wales, south of Ffestiniog, a mass
of solid rock as high as the Andes has been worn down
and carried bodily away; and that a few miles south
again, the peak of Arran Mowddy, which is now not two
thousand feet high, was once—either under the sea or
above it—nearer ten thousand feet.

If I am asked whither is all that enormous mass of
rock—millions of tons—gone? Where is it now? I know
not. But if I dared to hazard a guess, I should say it
went to make the New Red sandstones of England.

Surely, surely, there has been at least time enough
during all those ages for mud to change into slate.

And how was it changed?

Let us see first now, what was the origin of the mud.
The best answer I can give is to ask you to use your
imagination, or common sense; and to picture what
must go on in the case of a submarine eruption, such as
broke out off the coast of Iceland in 1788 and 1880, off

SNOWDON.

the Azores in 1811, and in our day in more than one spot in the Pacific Ocean.

A main bore or vent—or more than one—opens itself between the bottom of the sea and the nether fires. From each rushes an enormous jet of high-pressure steam and other gases, which boils up through the sea, and forms a cloud above; that cloud descends again in heavy rain, and gives out often true lightning from its under side.

But it does more. It acts as a true steam-gun, hurling into the air fragments of cold rock rasped off from the sides of the bore, and fragments also of melted lava, and clouds of dust, which fall again into the sea, and form there beds either of fine mud or of breccia—that is, fragments of stone embedded in paste.

This, the reader will understand, is no fancy sketch, as far as I am concerned. I have steamed into craters sawn through by the sea, and showing sections of beds of ash dipping outwards and under the sea, and in them boulders and pebbles of every size, which had been hurled out of the crater; and in them also veins of hardened lava, which had burrowed out through the soft ashes of the cone. Of those lava veins I will speak presently. What I want the reader to think of now is the immense quantity of ash which the steam-mitrailleuse hurls to so vast a height into the air, that it is often drifted many miles down to leeward.

To give two instances.

The jet of steam from Vesuvius, in the eruption of

1822, rose more than four miles into the air; the jet from the soufrière of St. Vincent in the West Indies, in 1812, probably rose higher; certainly it met the N.E. trade-wind, for it poured down a layer of ashes, several inches thick, not only on St. Vincent itself, but on Barbadoes, eighty miles to windward, and therefore on all the sea between.

Now let us consider what that fact represents. A layer of fine mud, laid down at the bottom of the ocean, several inches thick, eighty miles at least long, and twenty miles perhaps broad, and that, too, by one single erup-tion. Suppose that hardened in long ages (as it would be under pressure) into a bed of fine-grained Felstone, or volcanic ash; and we can understand how the ash-beds of Snowdonia—which may be traced some of them for many square miles—were laid down at the bottom of an ancient sea.

But to continue. First, as I said, the subterranean steam would blast a bore. The dust and stones rasped and blasted out of that hole would be spread about the sea-bottom as an ash-bed sloping away round the hole; then the molten lava would rise in the bore, and flow out over the ashes and the sea-bottom—perhaps in one direc-tion, perhaps all round. Then, usually, the volcano, having vented itself, would be quieter for a time, till the heat accu-mulated below, and more ash was blasted out, making a second ash-bed; and then would follow a second lava flow. Thus would be produced alternate beds of lava and ash, just as we find beds in slate-bearing mountains.

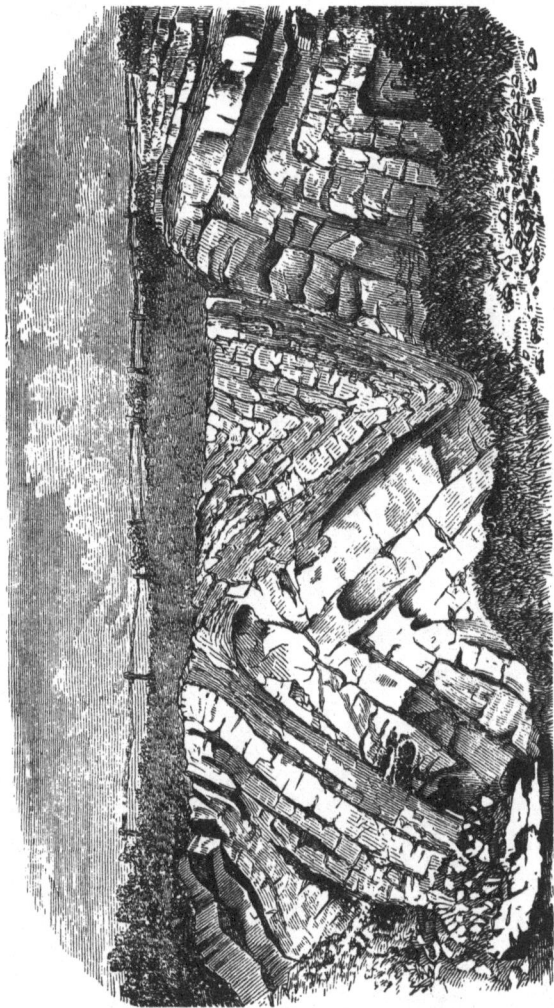

STONE BEDS, TWISTED AND TILTED.

That when the volcano has succeeded—as it did in the case of Sabrina Island off the Azores in 1811 ; in piling up an ash cone some thousand feet out of the sea ; that—as has happened to Sabrina Island—the cone is sunk again by earthquakes, and gnawn down at the same time by the sea-waves, till nothing is left but a shoal under water.

But where have all its vast heaps of ashes gone ?

To be spread about over the bottom of the sea, to mingle with the mud already there, and so make beds of which, like many in Snowdon, we cannot say whether they are of volcanic or of marine origin, because they are of both.

We now ask, naturally enough, how such a heap of beds, as I have described, can take the shape of mountains like Snowdon ?

Look at any sea cliff in which the strata are twisted and set on slope. There are hundreds of such in these isles. The beds must have been at one time straight and horizontal. But it is equally clear that they have been folded by being squeezed laterally. At least that is the simplest explanation, as may be proved by experiment. Take a number of pieces of cloth, or any such stuff; lay them on each other, and then squeeze them together at each end. They will arrange themselves in folds, just as the beds of the cliff have done. And if, instead of cloth, you take some more brittle matter, you will find that, as you squeeze on, these folds will tend to snap at the points of greatest tension or stretching, which will be of course at the anticlinal and synclinal lines—in plain English, the tops and bottoms of the folds. So cracks will be formed, and if the pressure goes on, the ends of the layers will shift against each other in the line of those cracks, forming faults like those so common in rocks.

But again, suppose that instead of squeezing these broken and folded lines together any more, you took off the pressure right and left, and pressed them upwards from below, by a mimic earthquake. They would rise; and as they rose leave open spaces between them. Now, if you could contrive to squeeze into them from below a paste, which would harden in the cracks and between the layers, and so keep them permanently apart, you would make them into a fair likeness of an average mountain range—a mess—if I may make use of a plain old word— of rocks which have, by alternate contraction and expansion, helped in the latter case by the injection of molten lava, been thrust about, as they are in most mountain ranges.

That such a contraction and expansion goes on in the crust of the earth is evident, for here are the palpable effects of it. And the simplest general cause which I can give for it is this: That things expand as they are heated, and contract as they are cooled.

Now, I am not learned enough—and were I, I have not time—to enter into the various theories which philosophers have put forward, to account for these grand phenomena.

The most remarkable, perhaps, and the most probable,

is the theory of M. Elie de Beaumont, which is, in a few words, this :—

That this earth, like all the planets, must have been once in a state of intense heat throughout, as its mass inside is probably now.

That it must be cooling, and giving off its heat into space.

That, therefore, as it cools, its crust must contract.

That, therefore, in contracting, wrinkles (for the loftiest mountain chains are nothing but tiny wrinkles, compared with the whole mass of the earth)—wrinkles, I say, must form on its surface from time to time. And that the mountain chains are these wrinkles.

Be that as it may, we may safely say this. That wherever the internal heat of the earth tends (as in the case of volcanoes) toward a particular spot, that spot must expand, and swell up, bulging the rocks out, and probably cracking them, and inserting melting lava into those cracks from below. On the other hand, if the internal heat leaves that spot again, and it cools, then it must contract more or less, in falling inward toward the centre of the earth ; and so the beds must be crumpled, and crushed, and shifted against each other still more, as those of our mountains have been.

Look at any piece of slate. All know that slate splits or cleaves freely in one direction only, into flat layers.

Now any one would suppose at first sight, and fairly enough, that the flat surface—the " plane of cleavage "— was also the plane of bedding. In simpler English we

should say, the mud which has hardened into this slate
was laid down horizontally. And therefore this slate is
one of the little horizontal beds of it, perhaps just what
was laid down in a single tide. We should have a right
to do so, because that would be true of most sedimentary

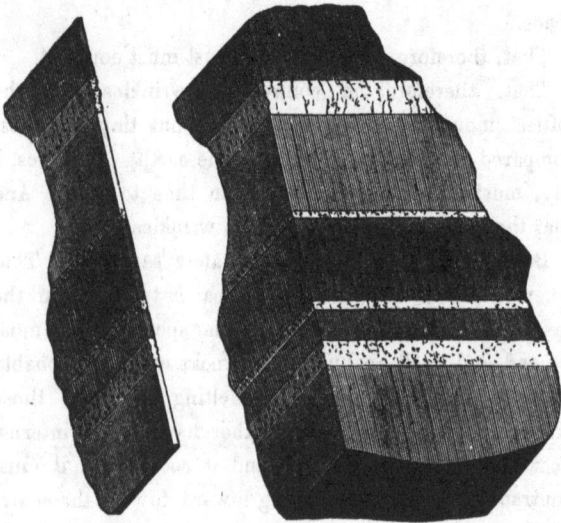

SLATE, SHOWING BED AND CLEVAGE.

rocks—*i.e.* rocks laid down by water. But it would not
be true of slate.

The plane of bedding in slate has nothing to do with
the plane of cleavage. Or, more plainly, the mud of which
the slate is made may have been deposited at the sea-
bottom and at right angles to the plane of cleavage.

These right-angled parallel planes of cleavage at each of which the slate splits away freely, will run through a whole mountain at the same angle.

Now what has made the slate rocks thus to split unlike all other water-laid rock, across the line of its deposit? We do not exactly know. All we can say is, that some force has gone on, or may be even now going on, in the more ancient rocks, which is similar to that which produces single crystals; some philosophers are of opinion that the force which has determined the cleavage of slates may be that of the electric currents, which (as is well known) run through the crust of the earth. Others, I believe, attribute the cleavage to the mere mechanical pressure of enormous weights of rock, especially where crushed by earthquakes. Others, again, point out that as these slates may have been highly heated, thermal electricity (*i.e.* electricity brought out by heat) may have acted on them.

One thing at least is clear. That the best slates are found among ancient lavas, which could not have happened without a proportionately enormous pressure, and therefore heat; and next that the best slates are invariably found in the oldest beds—that is, in the beds which have had most time to endure the changes, whether mechanical or chemical, which have made the earth's surface what we see it now.

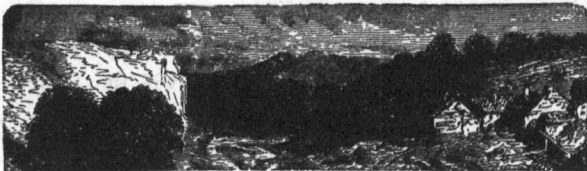

QUARRIES.

CHAPTER IV.

LIMESTONE.

I SHALL presume in all my readers some slight knowledge about lime. I shall take for granted, for instance, that all are better informed than a certain party of Australian black fellows were a few years since.

In prowling on the track of a party of English settlers, to see what they could pick up, they came—oh, joy!— on a sack of flour, dropped and left behind in the bush at a certain creek. The poor savages had not had such a prospect of a good meal for many a day. With endless jabbering and dancing, the whole tribe gathered round the precious flour-bag with all the pannikins, gourds, and other hollow articles it could muster, each of course with a due quantity of water from the creek therein, and the chief began dealing out the flour by handfuls, beginning, of course, with the boldest warriors. But, horror of horrors, each man's porridge swelled before his eyes, grew hot, smoked, boiled over. They turned and fled, man, woman, and child, from before that supernatural prodigy ; and the settlers coming back to look for the dropped sack, saw a sight which told the whole tale. For the poor creatures, in their terror, had thrown away their pans and calabashes, and each one filled with that which it was likely to contain, seeing that the sack itself had contained, not flour, but quick-lime. In memory of which comi-tragedy, that creek is called to this day, "Flour-bag Creek."

Lime is procured by boring the rock known as lime-stone, of which many of our hills are formed.

But now, what is the limestone ? and how did it get where it is ? Not into the mortar, I mean, but into the limestone quarry ?

Let me tell you, or rather, help you to tell yourselves, by leading you, as before, from the known to the un-known. Let me lead you to places unknown indeed to most. Let me lead you, in fancy, to some island in the Tropic seas. After all, I am not leading you as far away as you fancy by several thousand miles, as you will see, I trust, ere I have done.

Let me take you to some island : what shall it be like ? Shall it be a high island, with cliff piled on cliff, and peak on peak, all rich with mighty forests, like a furred mantle of green retreat, mounting up and up till it is

lost among white clouds above ?   Or shall it be a mere
low reef, which you do not see till you are close upon it;
on which nothing rises above the water, but here and
there a knot of cocoa-nut palms or a block of stone,
or a few bushes, all swarming with innumerable seafowl
and their eggs ?   Let it be which you will, both are
strange enough, both beautiful, both will tell us a
story.

The ship will have to lie-to, and anchor if she can,
it may be a mile, it may be only a few yards, from the
land.   For between it and the land will be a line of
breakers, raging in before the warm trade-wind.   And
this, you will be told, marks the edge of the coral reef.

You will have to go ashore in a boat, over a sea which
looks unfathomable, which may be a mile or more in
depth, and search for an opening in the reef, through
which the boat can pass without being knocked to
pieces.

You find one : and in a moment, what a change !
The deep has suddenly become shallow ; the blue, white,
from the gleam of the white coral at the bottom.   But
the coral is not all white, only indeed a little of it ;
for as you look down through the clear water, you find
that the coral is starred with innumerable live flowers,
blue, crimson, grey, every conceivable hue ; and that
these are the coral polypes, each with his ring of arms
thrust out of its cell, are building up their common
habitations of lime.

If you want to understand, by a rough but correct

CORAL ISLAND—AN ATOLL

description, what a coral polype is, all who have been to the seaside know, or at least have heard of, sea-anemones. Now coral polypes are sea-anemones, which make each a shell of lime, growing with its growth. As for their shapes, the variety of them, the beauty of them, no tongue can describe them.

If you want to see them, go to the Coral Rooms of the British or Liverpool Museums, and judge for yourselves. Only remember that you must re-clothe each of their exquisite forms with a coating of lime-jelly, of some delicate hue, and put back into every one of the thousand cells its living flower; and into the beds, or rather banks, of the salt-water flower-garden, the gaudiest of shell-less sea-anemones, such as we have on our coasts, rooted in the cracks, and live shells and sea-slugs, as gaudy as they, crawling about, with fifty other forms of fantastic and exuberant life. You must not overlook, too, the fish, especially the parrot-fish, some of them of the gaudiest colours, who spend their lives in browsing on the live coral, with strong clipping, and grinding teeth, just as a cow browses the grass, keeping the animal matter, and throwing away the lime in the form of an impalpable white mud, which fills up the interstices in the coral beds.

The bottom, just outside the reef, is covered with that mud, mixed with more lime-mud, which the surge wears off the reef; and if you have, as you should have, a dredge on board, and try a haul of that mud as you row home, you may find, but not always, animal form rooted in it, which will delight the soul of a scientific man. One I hope would be some sort of Terebratula, or shell akin to it. If you would think it a cockle, with forked beaks, you would be wrong. The animal which dwells in it has about the same relationship to a cockle as a dog has to a bird. It is a Brachiopod; a family with which the ancient seas once swarmed, but which is rare now, all over the world, having been supplanted and driven out of the seas by nearer and stronger forms

ANEMONE.

of shelled animals. The nearest spot at which you are likely to dredge a live Brachiopod would be in the deep water of Loch Fyne, in Argyleshire, where the species still linger, fastened strangely enough to the smooth pebbles of a submerged glacier, found in the open air during the age of ice, but sunk now to a depth of eighty fathoms. The first time I saw these shells come up in the dredge out of the dark and motionless abyss, I could sympathise with those feelings of mingled delight

and awe which, so my companion told me, the great
Professor Owen had in the same spot when he first
beheld these lingering remnants of a primæval world.

The other might be (but I cannot promise you even
a chance of dredging that, unless you were off the
coast of Portugal, or the windward side of some of

CORAL POLYPES.

the West Indian Islands) a live Crinoid, an exquisite
starfish, with long and branching arms, but rooted in
the mud by a long stalk, and that stalk throwing
out barren side branches, the whole a living plant of
stone. You may see in museums specimens of this
family, now so rare, all but extinct. And yet fifty

or a hundred different forms of the same type swarmed
in the ancient seas : whole masses of limestone made
up of little else but the fragments of such animals.

But we have not landed yet on the dry part of the
reef. Let us make to it, taking care meanwhile that
we do not get our feet cut by the coral, or stung as
by nettles by the coral insects. We shall see that the

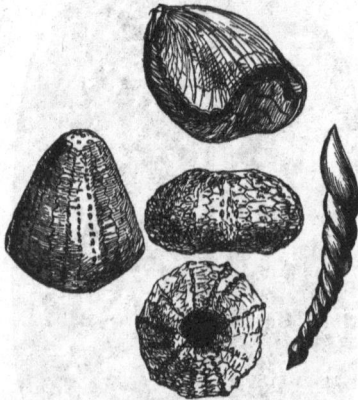

SHELLS OF CORAL MUD.

dry land is made up entirely of coral, ground and broken
by the waves, and hurled inland by the storm sometimes in huge boulders, mostly as fine mud; and that,
under the influence of the sun and of the rain, which
filters through it, charged with lime from the rotting
coral, the whole is setting, as cement sets, into rock.

And what is this ? A long bank of stone standing up
as a low cliff, ten or twelve feet above high-water mark.

It is full of fragments of shell, of fragments of coral, of all sorts of animal remains, and the lower part of it is quite hard rock ; moreover, it is bedded in regular layers just such as you see in a quarry. But how did it get there ? for it must have been formed at the sea-level, some of it, indeed, under the sea ; for here are great

STAR-FISH.

masses of madrepore and limestone corals unshedded just as they grew. What lifted it up ?

Your companions, if you have any, who know the island, have no difficulty in telling you. It was borne up, they say, in the earthquake in such and such a year ; and they will tell you, perhaps, that if you will go on shore, on the main island which rises inside the reef, you may see dead coral beds just like these, lying on the old rocks, and sloping up along the flanks of the mountains to several hundred feet above the sea. I have seen such many a time.

And so you find the coral being converted gradually into a limestone rock, either fine and homogeneous, composed of coral grown into pulp, or filled with corals and shells, or with angular fragments of older coral rock. Did you never see that last ? No ? Yes, you have a

CORAL ROCK.

hundred times. You have but to look at the marbles commonly used about these islands (perhaps for your own mantelpiece), with angular fragments embedded in the mass, and here and there a shell, and the whole cemented together by water, holding in solution car-bonate of lime, to see the very same phenomenon per-petuated to this day.

Thus, I think, we have got first from the known to the unknown, and from a tropic coral island back to the makers of our limestone hills ; and I did not speak at

random when I said that I was not leading you away
as far as you fancied by several thousand miles.

Examine any average limestone quarry from Bristol
to Berwick, and you will see there all that I have been
describing; that is, all of it which is not soft animal
matter, certain to decay. You will see the lime-mud
hardened into rock beds; you will see the shells em-

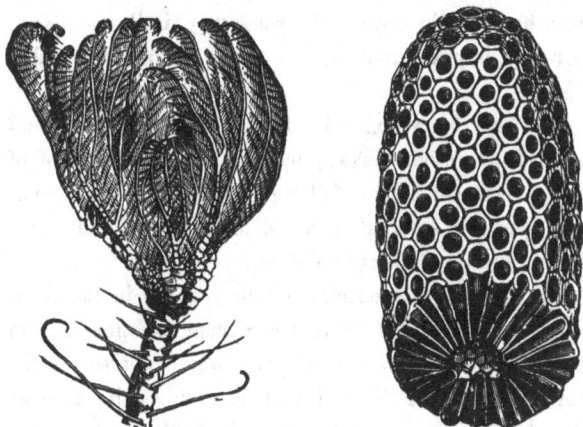

FORMS OF CORAL.

bedded in it; you will see the corals in every stage of
destruction; you will see whole layers made up of
innumerable fragments of Crinoids—no wonder they are
innumerable, for, it has been calculated, there are in a
single animal of some of the species 140,000 joints,
140,000 bits of lime to fall apart when its soft parts
decay. But was it not all there? And why should it

not have got there by the same process by which similar
old coral beds get up the mountain sides in the West
Indies and elsewhere—namely, by the upheaving force
of earthquakes? When you see similar effects, you have
a right to presume similar causes. If you see a man
fall off a house here, and break his neck, and some years
after, in London or New York, or anywhere else, you
find another man lying at the foot of another house,
with his neck broken in the same way, is it not a very
fair presumption that he has fallen off a house likewise?

You may be wrong. He may have come to his end
by a dozen other means; but you must have proof of
that. You will have a full right, in science and in common sense, to say that man fell off the house, till some
one proves to you that he did not.

In fact, there is nothing which you see in the limestones of these isles—save and except the difference in
every shell and coral—which you would not see in the
coral beds of the West Indies, if such earthquakes as
that famous one at St. Thomas, in 1866, became common and periodic, upheaving the land (they needs up-
heave it a very little, only two hundred and fifty feet),
till St. Thomas and all the Virgin Isles, and the mighty
mountain of Porto Rico, which looms up dim and purple
to the west, were all joined into dry land once more, and
the lonely coral-shoal of Anegada were raised, as it
would be raised then, into a limestone table-land, like
that of Central Ireland, of Galway, or of County Clare.

But you must clearly understand, that however much these coralline limestones have been upheaved *since they were formed*, yet the sea-bottom, *while they were being formed*, was sinking and not rising. This is a fact which was first pointed out by Mr. Darwin, in his observations made in the world-famous voyage of the *Beagle ;* and the observations of subsequent great naturalists have all gone to corroborate his theory.

It was supposed at first, you must understand, that when you found a coral island rising steeply to the sur-

SECTION OF ATOLL—showing sunken mountain, (dark) and coral formation (light).

face of the sea out of blue water, perhaps a thousand fathoms or more, that that was plain proof that the little coral polypes had begun at the bottom of the sea, and, in the course of ages, built up the whole island out of that enormous depth.

But it soon came out that that theory was not correct ; for the coral polypes cannot live and build save in shallow water—say thirty to forty fathoms. Indeed, some of the strongest and largest species work best at the very

surface, and in the cut of the strongest surf. And so
arose a puzzle as to how coral was formed at vast depths,
and of vast thickness. This Mr. Darwin explained. His
theory was—and there is no doubt now that it is correct—
that in these cases the sea-bottom is sinking ; that as it
sinks, carrying the coral beds down with it, the coral
dies, and a fresh live crop of polypes builds on the top
of the houses of their dead ancestors ; and as the depres-
sion goes on, generation after generation builds upwards,
the living on the dead, keeping the upper surface of the
reef at the same level, while its base is sinking downward
into the abyss.

And by applying this theory to the coral reef of the
Pacific Ocean, the following interesting facts were made
out :—

That where you have an island rising out of deep
water, with a ring of coral round it, a little way from
the shore—or, as in Eastern Australia, a coast with a
fringing reef (the Flinders reef of Australia is eleven
thousand miles long)—that is a pretty sure sign that
that shore, or mountain, is sinking slowly beneath the
sea. That where you have, as you often have in the
Pacific, a mere atoll, or circular reef of coral, with a
shallow pond of smooth water in the centre, and deep sea
round, that is a pretty sure sign that the mountain-
top has sunk completely into the sea, and that the
corals are going on building where its peak once was.
This latter condition is illustrated in the diagram.

And more. By working out the geography of the

South Sea Islands by the light of this theory of Darwin's, the following extraordinary fact has been discovered :—

That over a great part of the Pacific Ocean sinking is going on, and has been going on for ages ; and that the greater number of the beautiful and precious South Sea Islands are only the remnants of a vast continent or archipelago, which once stretched for thousands of miles between Australia and South America.

Now, applying the same theory to limestone beds, which are, as you know, only fossil coral reefs, we have a right to say, when we see in England, Scotland, and Ireland limestones several thousand feet thick, that while they were being laid down as coral reef, the sea-bottom, and probably the neighbouring land, must have been sinking to the amount of their thickness—to several thousand feet.

This is the story which, between the two, the limestone beds reveal.

CAVES

CAVES.

CHAPTER I.

HOW MADE.

D O you want to know all about the caves in which the old savages lived,—how they were made, and how the curious things inside them got there, and so forth?

Well, we will talk about that in good time; but now— What is that coming down the hill?

" Oh, only some chalk-carts."

Only some chalk-carts? It seems to me that these chalk-carts are the very things we want; that if we follow them far enough—I do not mean with our feet along the public road, but with our thoughts along a road which, I am sorry to say, the public do not yet know much about—we shall come to a cave, and understand how a cave is made.

Meanwhile, do not be in a hurry to say, " Only a chalk-

x

cart!" or only a mouse, or only a dead leaf. Chalk-carts, like mice, the dead leaves, and most other matters in the universe, are very curious and odd things in the eyes of wise and reasonable people. Whenever I hear young men saying "only" this and "only" that, I begin to suspect them of belonging, not to the noble army of martyrs, but to the ignoble army of noodles, who think nothing interesting or important but dinners, and balls, and races, and backbiting their neighbours; and I should be sorry to see you enlisting into that regiment when you grow up.

But think—are not chalk-carts very odd and curious things?

I think they are. To my mind, it is a curious question how men ever thought of inventing wheels; and, again, when they first thought of it. It is a curious question, too, how men ever found out that they could make horses work for them, and so began to tame them, instead of eating them; and a curious question (which I think we shall never get answered), when the first horse-tamer lived, and in what country. And a very curious, and, to me, beautiful sight it is, to see those two noble horses obeying that little boy, whom they could kill with a single kick.

But, beside all this, there is a question which ought to be a curious one to you (for I suspect you cannot answer it)—Why does the farmer take the trouble to send his cart and horses eight miles and more, to draw in chalk from Odiham chalk-pit?

"Oh, he is going to put it on the land, of course. They are chalking the bit at the top of the next field, where the copse was grubbed."

But what good will he do by putting chalk on it? Chalk is not rich and fertile, like manure. It is altogether poor, barren stuff: you know that, or ought to know it. Recollect the chalk cuttings and banks you have passed by on the railway—how utterly barren they are. Though they have been opened thirty years, not a blade of grass, hardly a bit of moss, has grown on them, or will grow, perhaps, for centuries.

Come, let us find out something about the chalk before we talk of the caves. The chalk is here, and the caves are not; and "Learn from the thing which lies nearest you," is as good a rule as "Do the duty which lies nearest you." Let us come into the grubbed bit, and ask the farmer—there he is in his gig.

"Well, old friend, and how are you? Here is a little boy who wants to know why you are putting chalk on your field?"

"Does he then? If he ever tries to farm round here, he will have to learn for his first rule—'No chalk, no wheat.'"

"But why?"

"Why, is more than I can tell, young squire. But if you want to see how it comes about, look here at this freshly-grubbed land—how sour it is. You can see that by the colour of it—some black, some red, some green, some yellow, all full of sour iron, which will let nothing grow:

After the chalk has been on it a year or two, those colours
will have all gone out of it; and it will turn to a nice
wholesome brown, like the rest of the field; and then
you will know that the land is sweet, and fit for any crop.
Now do you mind what I tell you, and then I'll tell you

CHALK-PIT.

something more. We put on the chalk because, beside
sweetening the land, it will hold water. You see, the
land about here, though it is often very wet from springs,
is sandy and hungry; and when we drain the bottom

water out of it, the top water (that is, the rain) is apt to run through it too fast, and then it dries and burns up; and we get no plant of wheat, nor of turnips either. So we put on chalk to hold water, and keep the ground moist."

"But how can these lumps of chalk hold water? They are not made like cups."

"No; but they are made like sponges, which serves our turn better still. Just take up that lump, young

CAVERN.

squire, and you'll see water enough in it, or rather looking out of it, and staring you in the face."

"Why! one side of the lump is all over thick ice."

"So it is. All that water was inside the chalk last night, till it froze. And then it came squeezing out of the holes in the chalk in strings, as you may see it if you break the ice across. Now you may judge for yourself how much water a load of chalk will hold, even on a dry

summer's day. And now, if you'll excuse me, sir, I must
be off to market."

"Was it all true that the farmer said ?"

Quite true, I believe. He is not a scientific man—
that is, he does not know the chemical causes of all these
things ; but his knowledge is sound and useful, because it
comes from long experience.

What he meant by the chalk sweetening the land you
would not understand yet, and I can hardly tell you ; for
chemists are not yet agreed how it happens. But he was
right ; and right, too, what he told you about the water
inside the chalk, which is more important to us just now ;
for, if we follow it out, we shall surely come to a cave at
last.

So now for the water in the chalk. You can see now
why the chalk-downs at Winchester are always green,
even in the hottest summer ;—because "Madam How" has
put under them her great chalk sponge. The winter rains
soak into it ; and the summer heat draws that rain out of
it again as invisible steam, coming up from below, to
keep the roots of the turf cool and moist under the blazing
sun.

But all the hollows of the chalk-downs are as dry as
the hill-tops. There is not a brook, or the mark of a
water-course, in one of them. You are like the Ancient
Mariner in the poem, with

> " Water, water everywhere,
>   Yet not a drop to drink."

To get that you must go down and down, hundreds of

feet, to the green meadows through which silver Itchen glides toward the sea. There you stand upon the bridge, and watch the trout in water so crystal-clear that you see every weed and pebble as if you looked through air. If ever there was pure water, you think, that is pure. Is it so? Drink some. Wash your hands in it and try. You feel that the water is rough, hard (as they call it), quite different from the water at home, which feels as soft as velvet. What makes it so hard?

Because it is full of invisible chalk. In every gallon of that water there are, perhaps, fifteen grains of solid chalk, which was once inside the heart of the hills above. Day and night, year after year, the chalk goes down to the sea; and if there were such creatures as water-fairies—if it were true, as the old Greeks and Romans thought, that rivers were living things, with a Nymph who dwelt in each of them, and was its goddess or its queen,—then, if your ears were open to hear her, the Nymph of Itchen might say to you—

" So you think that I do nothing, but, as Mr. Tennyson's beautiful song says,—

> ' Chatter over stony ways,
>   In little sharps and trebles ;
> And bubble into eddying bays,
>   And babble on the pebbles.'

" Yes ; I do that : and I love to tell how

> ' I wind about, and in and out,
>   With here a blossom sailing,
> And here and there a lusty trout,
>   And here and there a grayling,

' And here and there a foamy flake
　Upon me, as I travel
With many a silvery waterbreak
　Above the golden gravel,

' And draw them all along, and flow
　To join the brimming river,
For men may come and men may go,
　But I go on for ever.'

"Yes; that is all true; I only exist (like everything else, from the sun in heaven to the gnat which dances in his beam) on condition of working, whether we wish it or not, whether we know it or not. So day by day, and night by night, while you are sleeping (for I never sleep), I carry, delicate and soft as I am, a burden which giants could not bear: and yet I am never tired. Every drop of rain which the south-west wind brings from the West Indian seas gives me fresh life and strength to bear my burden: and it has need to do so; for every drop of rain melts a tiny grain of chalk, and helps to send it down through the solid hill by one of the million pores and veins which at once feed and burden my springs. Ages on ages I have worked on thus, carrying the chalk into the sea. And ages on ages, it may be, I shall work on yet; till I have done my work at last and levelled the high downs into a flat sea-shore, with beds of flint gravel rattling in the shallow waves."

She might tell you that; and when she had told you, you would surely think of the clumsy chalk-cart rumbling down the hill, and then of the graceful stream, bearing silently its invisible load of chalk; and see how much more delicate and beautiful, as well as vast and wonderful, is Nature's work than man's.

But why does the stream thus work on for ever?

It is for a reasonable being to guess why; or at least to listen to me if I guess for you, and say, perhaps—I only say perhaps—that chalk may be going to make layers of rich marl in the sea between England and France; and those marl-beds may be upheaved and grow into dry land, and be ploughed, and sowed, and reaped by a wiser race of men, in a better-ordered world than this; or the chalk may have even a nobler destiny before it. That may happen to it, which has happened already to many a grain of lime. It may be carried thousands of miles away to help in building up a coral reef. That coral reef may harden into limestone beds. Those beds may be covered up, pressed and, it may be, heated, till they crystallize into white marble; and out of it fairer statues be carved, and grander temples built, than the world has ever yet seen.

And if that is not the reason why the chalk is being sent into the sea, then there is another reason, and probably a far better one.

But you will say now, that we have followed the chalk-cart a long way without coming to the cave. You are wrong.

CAVES

CHAPTER II.

HOW MADE.—(*Continued.*)

WE have come now to the very threshold of the cave. And if we follow on we shall walk into the corridors and caverns and halls and behold rivers and cascades underground, stalactite pillars and stalagmite statues, and all the wonders of the grottoes of Adelsberg, Antiparos, or Kentucky.

"But there are no caves in chalk?"

No, not that I ever heard of. There are, though, in limestone, which is only a harder kind of chalk. And this chalk could be turned into hard limestone, I believe, even now; and in more ways than one; but in ways which would not be very comfortable or profitable for Southern folk who live on it. I am afraid that—what between squeezing and heating—she would flatten them all out into phosphatic fossils, about an inch thick; and turn Winchester city into a " breccia," which would puzzle geologists a hundred thousand years hence. So we will hope that she will leave our chalk-downs for the Itchen to wash gently away, while we talk about caves, and how they are scooped out by water underground just

STALAGMITES AND STALACTITES.

in the same way, only more roughly, as the chalk is melted.

Suppose, then, that these hills, instead of being soft, spongy chalk, were all hard limestone marble. Then the rain-water, instead of sinking through the chalk as now, would run over the ground downhill, and if it came to a crack (a fault, as it is called), it would run down between the rock; and as it ran it would eat that hole wider and wider year by year, and make a " swallow hole "—such as you may see in plenty if you ever go up Whernside, or any of the high hills in Yorkshire—unfathomable pits in the green turf, in which you may hear the water tinkling and trickling far, far underground.

And now, before we go a step farther, you may under-stand why the bones of animals are so often found in limestone caves.   Down such swallow-holes how many beasts must fall : either in hurry and fright, when hunted by lions and bears and such cruel beasts; or more often still in the time of snow, when the holes are covered with drift; or, again, if they died on the open hillsides, their bones might be washed in, in floods, along with mud and stones, and buried with them in the cave below; and besides that, lions and bears and hyænas might live in the caves below, as we know they did in some caves, and drag in bones through the caves' mouths; or, again, savages might live in that cave, and bring in animals to eat, like the wild beasts; and so those bones might be mixed up, as we know they were, with things which the savages had left behind—like flint tools or beads; and then the whole would be hardened by the dripping of the limestone water, into a paste of breccia just like this in my drawer.   But the bones of the savages themselves you would seldom or never find mixed in it—unless some one had fallen in by accident from above.

CAVE OF ANTIPAROS.

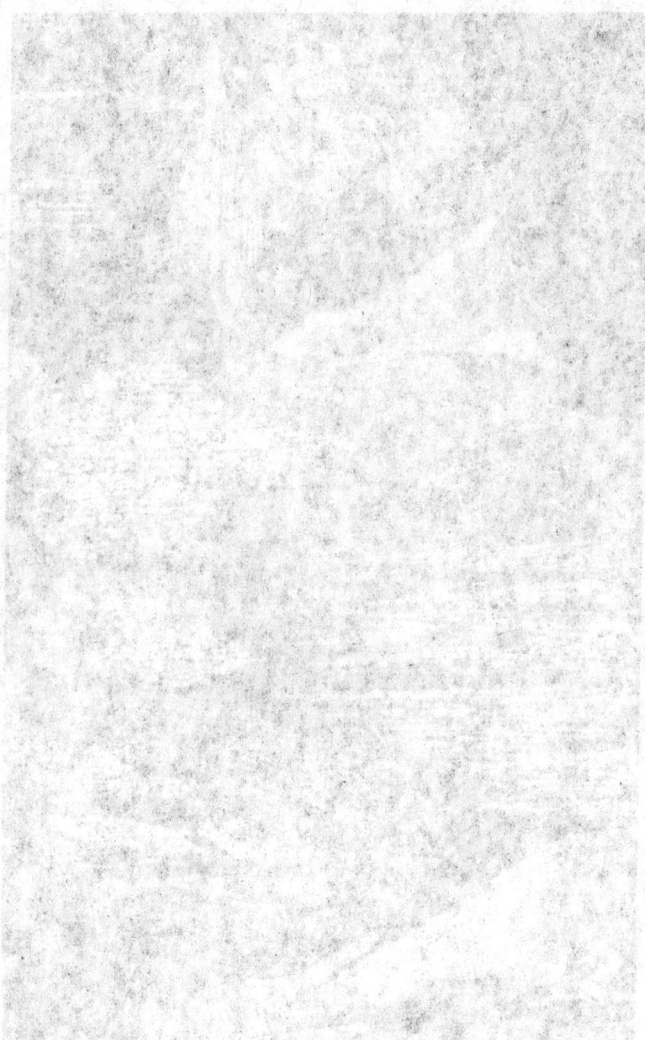

And why ?   (For there is a Why ? to that question.)

Simply because they were men ; and because God has put into the hearts of all men, even of the lowest savages, some sort of reverence for those who are gone ; and has taught them to bury, or in some other way take care of, their bones.

But how is the swallow-hole sure to end in a cave ?

Because it cannot help making a cave for itself if it has time.

Think ; and you will see that it must be so.   For that water must run somewhere ; and so it eats its way out between the beds of the rock, making underground galleries, and at last caves and lofty halls.   For it always eats, remember, at the bottom of its channel, leaving the roof alone.   So it eats, and eats more in some places and less in others, according as the stone is harder or softer, and according to the different direction of the rock-beds (what we call their dip and strike) ; till at last it makes one of those wonderful caverns about which you are so fond of reading,—such a cave as there actually is in the rocks of the mountain of Whernside, fed by swallow-holes around the mountain-top ; a cave hundreds of yards long, with halls, and lakes, and waterfalls, and curtains and festoons of stalactite which have dripped from the roof, and pillars of stalagmite which have been built up on the floor below.

These stalactites—among the most beautiful of all Nature's work—are sometimes like branches of roses or of grapes ; sometimes like statues ; sometimes like deli-

cate curtains, and I know not what other beautiful
shapes. But they are all made in the same way ; just in
the same way as those little straight stalactites which
you may have seen hanging, like icicles, in vaulted
cellars, or under the arches of a bridge. The water melts
more lime than it can carry, and drops some of it again,
making fresh limestone grain by grain as it drips from
the roof above ; and fresh limestone again where it
splashes on the floor below: till, if it dripped long
enough, the stalactite hanging from above would meet
the stalagmite rising from below, and join in one straight
round white graceful shaft, which would seem (but only
seem) to support the roof of the cave. And out of that
cave—though not always out of the mouth of it—will run
a stream of water, which seems to you clear as crystal,
though it is actually, like the Itchen at Winchester, full
of lime ; so full of lime, that it makes beds of fresh
limestone, which are called Travertine—which you may
see in Italy, and Greece, and Asia Minor ; or perhaps it
petrifies, as you call it, the weeds in its bed, like that
dropping-well at Knaresborough.

And the cause is this. The water is so full of lime,
that it is forced to throw away some of it upon everything
it touches, and so encrusts with stone—though it does
not turn to stone—almost anything you put in it. You
have seen, or ought to have seen, petrified moss and
birds' nests and such things from Knaresborough Well :
and now you know a little, though only a very little, of
how the pretty toys are made.

Now if you can imagine for yourself the amount of lime which one of these subterranean rivers would carry away, gnawing underground day and night, summer and winter, centuries after centuries, then you will not be surprised at the enormous size of caverns which may be seen in different parts of the world; but always, I believe, in limestone rock.

You would not be surprised (though you would admire them) at the caverns of Adelsberg in Carniola (in the South of Austria, near the top of the Adriatic), which run, I believe, for miles in length; and in the lakes of which, in darkness from its birth until its death, lives that strange beast the Proteus, a sort of long newt which never comes to perfection; I suppose for want of the genial sunlight which makes all things grow. But he is blind; and more, he keeps all his life the same feathery gills which newts have when they are babies, and which we have so often looked at through the microscope, to see the blood-globules run round and round inside.

You would not wonder, either, at the Czirknitz Lake, near the same place, which at certain times of the year vanishes suddenly through chasms under water, sucking the fish down with it; and after a certain time boils suddenly up again from the depths, bringing back with it the fish, who have been swimming comfortably all the time in a subterranean lake; and bringing back, too (and, extraordinary as this story is, there is good reason to believe it true), live wild ducks who went down small and unfledged, and come back full-grown and fat, with water,

Y

weeds, and small fish in their stomachs, showing they
have had plenty to feed on underground. But—and this
is the strangest part of the story, if true—they come up
unfledged just as they went down, and are moreover
blind from having been so long in darkness. After a

CAVERN AVENUE.

while, however, folks say, their eyes get right, their
feathers grow, and they fly away like other birds.

Neither would you be surprised at that Mammoth Cave,
in Kentucky, the largest cave in the known world,

through which you may walk nearly ten miles on end ;
and in which a hundred miles of gallery have been explored already, and yet no end found to the cave.   In
it (the guides will tell you) there are " two hundred and
twenty-six avenues, forty-seven domes, eight cataracts,

CAVERN LAKE.

twenty-three pits, and several rivers."   And if that fact is
not very interesting to you, I will tell you something
which ought to interest you: that this cave is so immensely old that various kinds of little animals, who have
settled themselves in the outer parts of it, have had time

to change their shape, and to become quite blind; so that blind fathers and mothers have blind children, generation after generation.

There are blind rats there, with large shining eyes which cannot see; blind land-crabs, who have the foot-stalks of their eyes (you may see them in any crab) still left; but the eyes which should be on the top of them are gone. There are blind fish, too, in the cave, and blind insects; for, if they have no use for their eyes in the dark, why should "Madam How" take the trouble to finish them off?

One more cave I must tell you of, to show you how old some caves must be, and then I must stop; and that is the cave of Caripé, in Venezuela, which is the most northerly part of South America. There, in the face of a limestone cliff, crested with enormous flowering trees, and festooned with those lovely creepers of which you have seen a few small ones in hot-houses, there opens an arch which runs straight in like a cathedral nave for more than fourteen hundred feet. Out of it runs a stream; and along the banks of that stream, as far as the sunlight strikes in, grow wild bananas, and palms, and "lords and ladies" (as you call them), which are not, like ours, one foot, but many feet high. Beyond that the cave goes on, with subterranean streams, cascades, and halls, no man yet knows how far.

A friend of mine last year went in farther, I believe, than any one yet has gone; but, instead of taking Indian torches made of bark and resin, or even torches made of Spanish wax, such as a brave bishop of those parts used once when he went in farther than any one before him, he took with him some of that beautiful magnesium light which you have seen often here at home. And in one place, when he lighted up the magnesium, he found himself in a hall full three hundred feet high—higher far, that is, than the dome of St. Paul's—and a very solemn thought it was to him, he said, that he had seen what no other human being ever had seen; and that no ray of light had ever struck on that stupendous roof in all the ages since the making of the world.

But if he found out something which he did not expect, he was disappointed in something which he did expect. For the Indians warned him of a hole in the floor which (they told him) was an unfathomable abyss. And lo, and behold! when he turned the magnesium light upon it, the said abyss was just about eight feet deep. But it is no wonder that the poor Indians with their little smoky torches should make such mistakes; no wonder, too, that they should be afraid to enter far into those gloomy vaults; that they should believe that the souls of their ancestors live in that dark cave; and that they should say that when they die they will go to the Guacharos, as they call the birds that fly with doleful screams out of the cave to feed at night, and in again at daylight, to roost and sleep.

Now, it is these very Guacharo birds which are to me the most wonderful part of the story. The Indians kill and eat them for their fat, although they believe they

have to do with evil spirits. But scientific men who
have studied these birds will tell you that they are more
wonderful than if all the Indians' fancies about them were
true. They are great birds, more than three feet across
the wings, somewhat like owls, somewhat like cuckoos,
somewhat like goatsuckers; but, on the whole, unlike
anything in the world but themselves; and instead of
feeding on moths or mice, they feed upon hard dry fruits,
which they pick off the trees after the setting of the sun.
And wise men will tell you, that the making of such a
bird as that, and giving it that peculiar way of life, and
settling it in that cavern, and a few more caverns in that
part of the world, and therefore the making the caverns
ready for them to live in, must have taken ages and ages,
more than you can imagine or count.

CAVES.

CHAPTER III.

WHAT THEY CONTAIN.

LOOK at this and judge for yourself.

What is it? a piece of old mortar? Yes. But mortar which was made by Nature, and not by any man.

And what is in it?

A piece of flint and some bits of bone.

But look at that piece of flint. It is narrow, thin, sharp-edged: quite different in shape from any bit of flint which you or I ever saw among the hundreds of thousands of broken bits of gravel which we tread on here all day long.

And here are some more bits like it, which came from the same place—all very much the same shape, like rough knives or razor blades; and here is a core of flint, the remaining part of a large flint, from which, as you may

see, blades like those have been split off. Those flakes of
flint were split off by men ; even young eyes ought to be
able to see that.

And here are other pieces of flint—pear-shaped, but
flattened, sharp at one end and left rounded at the other,
which look like spear-heads, or arrow-heads, or pointed
axes, or pointed hatchets—even your young eyes can see
that these must have been made by man. And they are,
I may tell you, just like the tools of flint, or of obsidian,
which is volcanic glass, and which savages use still where
they have not iron. There is a great obsidian knife, you
know, in this very parish, which came from Mexico ; and
your eye can tell you how like it is to these flint ones.

But these flint tools are very old.

If you crack a fresh flint, you will see that its surface
is grey, and somewhat rough, so that it sticks to your
tongue. These tools are smooth and shiny ; and the
edges of some of them are a little rubbed from being
washed about in gravel ; while the iron in the gravel has
stained them reddish, which it would take hundreds and
perhaps thousands of years to do.

There are little rough markings, too, upon some of
them, which, if you look at through a magnifying
glass, are iron, crystallized into the shape of little sea-
weeds and trees—another sign that they are very very
old.

And what is more, near the place where these flint
flakes come from there are no flints in the ground for
hundreds of miles ; so that men must have brought them

there ages and ages since. And to tell you plainly, these are scrapers, such as the Esquimaux in North America still use to scrape flesh off bones, and to clean the insides of skins.

But did these people (savages perhaps) live when the country was icy cold?

FLINT TOOLS.

Look at the bits of bone. They have been split, you see, lengthways; that, I suppose, was to suck the marrow out of them, as savages do still. But to what animal do the bones belong?

That is the question; and one which I could not have

answered you, if wiser men than I am could not have told me.

They are the bones of reindeer—such reindeer as are now found only in Lapland, and the half-frozen parts of North America, close to the Arctic circle, where they have six months day and six months night. You have read of Laplanders, and how they drive reindeer in their sledges, and live upon reindeer milk; and you have read of Esquimaux, who hunt seals and walruses, and live in houses of ice, lighted by lamps fed with the same blubber on which they feed themselves. I need not tell you about them.

Now comes the question—Whence did these flints and bones come?

They came out of a cave in Dordogne, in the heart of sunny France,—far away to the south where it is hotter every summer than it was here even this summer, from among woods of box and evergreen, oak, and vineyards of rich red wine. In that warm land once lived savages, who hunted amid ice and snow the reindeer, and with the reindeer animals stranger still.

And now I will tell you a fairy tale: to make you understand it at all I must put it in the shape of a tale. I call it a fairy tale, because it is so strange; indeed, I think I ought to call it the fairy tale of all fairy tales, for by the time we get to the end of it I think it will explain to you how our forefathers got to believe in fairies, and trolls, and elves, and scratlings, and all strange little people who were said to haunt the mountains and the caves.

BANK OF ENGLISH RIVER, LONG AGO.

Well, once upon a time, so long ago that no man can tell when, the land was so much higher, that between England and Ireland, and, what is more, between England and Norway, was firm dry land.

The country then must have looked—at least we know it looked so in Norfolk—very like what our moors look like here. There were forests of Scotch fir, and of spruce, too, which is not wild in England now, though you may see plenty in every plantation.   There were oaks and alders, yews and sloes, just as there are in our woods now.   There was buck-bean in the bogs, as there is in Larmer's and Heath pond; and white and yellow water-lilies, horn-wort, and pond-weeds, just as there are now in our ponds.   There were wild horses, wild deer, and wild oxen, those last of an enormous size.   There were little yellow roe-deer, which will not surprise you, for there are hundreds and thousands in Scotland to this day; and, as you know, they will thrive well enough in our woods now.   There were beavers too; but that must not surprise you, for there were beavers in South Wales long after the Norman Conquest, and there are beavers still in the mountain glens of the south-east of France.   There were honest little water-rats too, who I daresay sat up on their hind legs like monkeys, nibbling the water-lily pods, thousands of years ago, as they do in our ponds now.

Well, so far we have come to nothing strange; but now begins the fairy tale.   Mixed with all these animals, there wandered about great herds of elephants and rhinoceroses; not smooth-skinned, mind, but covered with hair and wool, like those which are still found sticking out of the everlasting ice-cliffs, at the mouth of the Lena and other Siberian rivers, with the flesh, and skin, and hair so fresh upon them, that the wild wolves

COAST GLACIER.

tear it off, and snarl and growl over the carcase of monsters who were frozen up thousands of years ago. And with them, still stranger far, were great hippopotami; who came, perhaps, northward in summer time

along the sea-shore and down the rivers, having spread
hither all the way from Africa; for in those days, you
must understand, Sicily, and Italy, and Malta—look at
your map—were joined to the coast of Africa: and so it

SIBERIAN MAMMOTH.

may be was the rock of Gibraltar itself; and over the sea
where the Straits of Gibraltar now flow was firm, dry
land, over which hyænas and leopards, elephants and
rhinoceroses ranged into Spain, for their bones are found

at this day in the Gibraltar caves.   And this is the first
chapter of my fairy tale.

Now, while all this was going on, and perhaps before
this began, the climate was getting colder and colder
year by year—we do not know how; and what is more,
the land was sinking; and it sank so deep that at last
nothing was left out of the water but the tops of the
mountains in Ireland, and Scotland, and Wales. It sank
so deep, that it left beds of shells belonging to the Arctic
regions nearly two thousand feet high upon the mountain side.   And so

> "It grew wondrous cold,
> And ice, mast-high, came floating by
> As green as emerald."

But there were no masts then to measure the icebergs
by, nor any ship nor human being there.   All we know
is that the icebergs brought with them vast quantities of
mud, which sank to the bottom, and covered up that
pleasant old forest-land in what is called boulder clay;
clay full of bits of broken rock, and of blocks of stone so
enormous, that nothing but an iceberg could have carried
them.

So all the animals were drowned or driven away, and
nothing was left alive perhaps, except a few little hardy
plants which clung about cracks and gullies in the mountain tops; and whose descendants live there still.   That
was a dreadful time; the worst, perhaps, of all the age of
ice; and so ends the second chapter of my fairy tale.
Now for my third.

"When things come to the worst," says the proverb, "they commonly mend;" and so did this poor frozen and drowned land of England and France and Germany, though it mended very slowly. The land began to rise once more, and rose till it was perhaps as high as it had been at first, and hundreds of feet higher than it is now; but still it was very cold, covered, in Scotland at least, with one great sea of ice and glaciers descending down into the sea.

CAVES.

CHAPTER IV.

WHAT THEY CONTAIN.—(*Continued.*)

A S the land rose, and grew warmer, too, while it rose, the wild beasts who had been driven out by the great drowning came gradually back again.

The bottom of the old icy sea turned into dry land, and got covered with grasses and shrubs once more. Elephants, rhinoceroses, hippopotami, oxen—sometimes the same species, sometimes slightly different—returned to France, and then to England (for there was no British Channel then to stop them); and with them came other strange animals, the great Irish elk, as he is called, as large as the largest horse, with horns sometimes fifteen feet across. A pair of those horns with the skull you have seen yourself, and can judge what a noble animal he must have been. Enormous bears came too, and hyænas, and a tiger or lion (I cannot say which), as large as the largest Bengal tiger now to be seen in India.

And in those days—we cannot, of course, exactly say when—there came—first, I suppose, into the south and east of France, and then gradually onward into England and Scotland and Ireland—creatures without any hair to keep them warm, or scales to defend them, without horns or tusks to fight with, or teeth to worry and bite; the weakest, you would have thought, of all the beasts, and yet stronger than all the animals, because they were Men, with reasonable souls.

Whence they came we cannot tell, nor why; perhaps from mere hunting after food, and love of wandering and being independent and alone. Perhaps they came into that icy land for fear of stronger and cleverer people than themselves; for we have no proof, as is often taken for granted, that they were the first men that trod this earth. But be that as it may, they came; and so cunning were these savage men, and so brave likewise, though they had no iron among them, only flint and sharpened bones, yet they contrived to kill and eat the mammoths, and the giant oxen, and the wild horses, and the reindeer, and to hold their own against the hyænas, and tigers, and bears, simply because they had wits, and the dumb animals had none.

And that is the strangest part to me of all my fairy tale. For what a man's wits are, and why he has them, and therefore is able to invent and to improve, while even the cleverest ape has none, and therefore can invent and improve nothing, and therefore cannot better himself, but must remain from father to son, and father to son again, a stupid, pitiful, ridiculous ape, while men can go on civilising themselves, and growing richer and more comfortable, wiser and happier, year by year,—how that

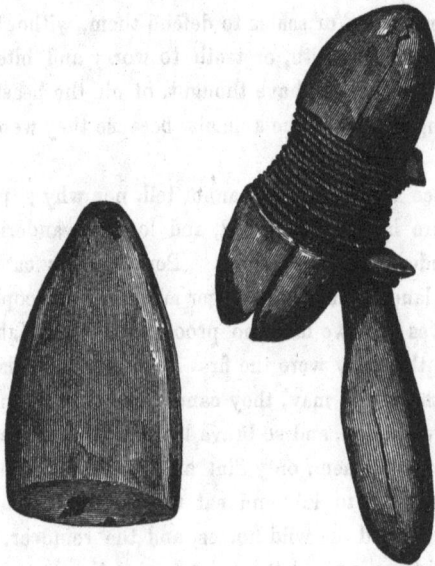

STONE TOOLS.

comes to pass, I say, is to me a wonder and a prodigy and a miracle, stranger than all the most fantastic marvels you ever read in fairy. tales.

You may find the flint weapons which these old savages used buried in many a gravel-pit up and down

France and the south of England; in caves which water has eaten out of the limestone rocks, like that famous cave of Kent's Hole at Torquay.

In it, and in many another cave, lie the bones of animals which the savages ate, and cracked to get the

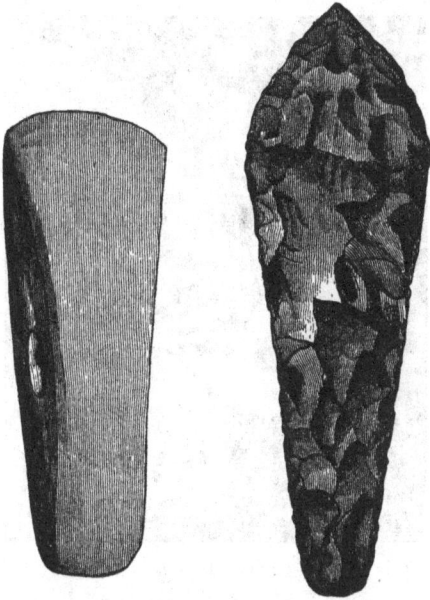

STONE TOOLS.

marrow out of them, mixed up with their flint-weapons and bone harpoons, and sometimes with burnt ashes and with round stones, used perhaps to heat water, as savages do now, all baked together into a hard paste or breccia by the lime. These are in the water, and are often covered with a floor of stalagmite which has dripped from the roof above and hardened into stone.

In these caves, no doubt, the savages lived; for not only have weapons been found in them, but actually

CAVE SAVAGE.

drawings scratched (I suppose with flint) on bone or mammoth ivory; drawings of elk, and bull, and horse, and ibex; and one, which was found in France, of the great mammoth himself, the woolly elephant, with a mane on his shoulders like a lion's mane. So you see

that one of the earliest fancies of this strange creature
called man was to draw, as you and your schoolfellows
love to draw, and copy what you see, you know not
why.

Remember that. You like to draw; but why you like
it neither you nor any man can tell. It is one of the

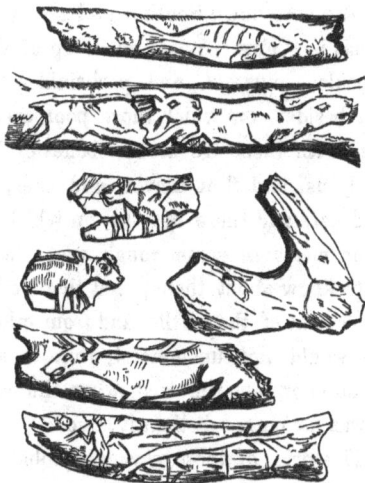

BONE CARVINGS.

mysteries of human nature; and that poor savage clothed
in skins, dirty it may be, and more ignorant than you
(happily) can conceive, when he sat scratching on ivory
in the cave the figures of the animals he hunted, was
proving thereby that he had the same wonderful and

mysterious human nature as you; that he was the kins-
man of every painter and sculptor who ever felt it a
delight and duty to copy the beautiful works of God.

Sometimes, again, especially in Denmark, these savages
have left behind upon the shore mounds of dirt, which
are called there "kjökken-möddings," "kitchen-middens,"
as they would say in Scotland, "kitchen dirt-heaps," as
we should say here down South; and a very good name
for them that is. For they are made up of the shells of
oysters, cockles, mussels, and periwinkles, and other
shore-shells besides, on which those poor creatures fed;
and mingled with them are broken bones of beasts, and
fishes, and birds, and flint knives, and axes, and sling-
stones; and here and there hearths, on which they have
cooked their meals in some rough way; and that is
nearly all we know about them; but this we know from
the size of certain of the shells, and from other reasons
which you would not understand, that these mounds
were made an enormous time ago, when the water of the
Baltic Sea was far more salt than it is now.

Out of old stories of these savages probably grew up
the stories of fairies, elves, and trolls, and scratlings, and
cluricaunes, and ogres, of which you have read so many.

When stronger and bolder people, like the Irish, and
the Highlanders of Scotland, and the Gauls of France,
came northward with their bronze and iron weapons;
and still more when our own forefathers, the Germans
and the Norsemen, came, these poor little savages, with
their flint arrows and axes, were no match for them. and

had to run away northward, or be all killed out; for people were fierce and cruel in those old times, and looked on every one of a different race from themselves as a natural enemy.

They had not learnt—alas! too many have not learnt it yet—that all men are brothers for the sake of Jesus Christ our Lord. So these poor savages were driven out, till none were left, save the little Lapps up in the north of Norway, where they live to this day.

But stories of them, and of how they dwelt in caves and had strange customs, and used poisoned weapons, and how the elf bolts (as their flint arrow-heads are still called) belonged to them, lingered on, and were told round the fire on winter nights, and added to, and played with half in fun, till a hundred legends sprang up about them, which used once to be believed by grown-up folk, but which now only amuse children.

Because some of these savages were very short, as the Lapps and Esquimaux are now, the story grew of their being so small that they could make themselves invisible; and because others of them were (but probably only a few) very tall and terrible, the story grew that there were giants in that old world, like that famous Gogmagog, whom Brutus and his Britons met (so old fables tell), when they landed first at Plymouth; and fought him, and threw him over the cliff.

Ogres, too—of whom you read in fairy tales—I am afraid that there were such people once, even here in Europe; strong and terrible savages who ate human beings. Of course, the legends and tales about them became ridiculous and exaggerated as they passed from mouth to mouth over the Christmas fire, in the days when no one could read or write. But that the tales began by being true any one may well believe who knows how many cannibal savages there are in the world even now.

I think that, if ever there was an ogre in the world, he must have been very like a certain person who lived, or was buried, in a cave in the Wanderthal, between Elberfeld and Düsseldorf, on the Lower Rhine. The skull and bones which were found there (and which are very famous now among scientific men) belonged to a personage whom I should have been very sorry to meet in the wild forest; to a savage of enormous strength of limb (and I suppose of jaw), likewise

> "like an ape,
> With forehead villanous low;"

who could have eaten you if he would; and (I fear) also would have eaten you if he could.

Such savages may have lingered (I believe, from the old ballads and romances, that they did linger) for a long time in lonely forests and mountain caves, till they were all killed out by warriors who wore mail-armour, and carried steel sword, and battle-axe, and lance.

So perhaps truth lies behind the fairy fictions, and if so, what is behind is more wonderful than the fiction. Even a child like you will see that it must be so, if you will but recollect who makes fiction, and who makes facts.

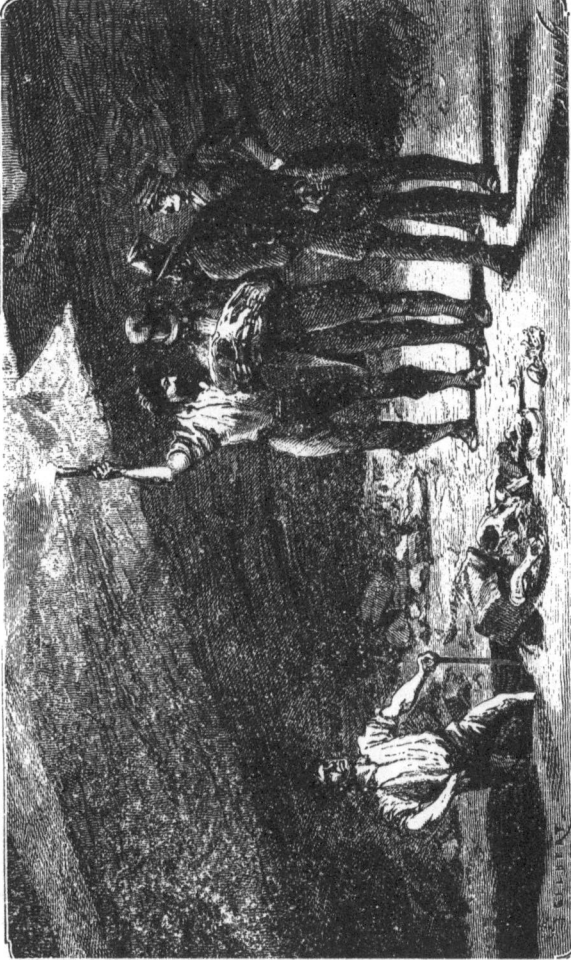

FINDING A SKULL.

Man makes fiction: he invents stories pretty enough, fantastical enough.

But out of what does he make them up? Out of a few things in this great world which he has seen, and heard, and felt, just as he makes up his dreams.

But who makes truth? Who makes facts? Who, but God?

Then truth is as much larger than fiction, as God is greater than man; as much larger as the whole universe is larger than the little corner of it that any man, even the greatest poet or philosopher, can see. And as much grander, and as much more beautiful, and as much more strange. For one is the whole, and the other is a few tiny scraps of the whole. The one is the work of man's mind; the other is the work of God.

Be sure that no man can ever fancy anything strange, unexpected, and curious, without finding, if he had eyes to see, a hundred things around his feet more strange, more unexpected, more curious, actually ready-made already by God.

Is it not so, then, that novels and story-books are scarcely worth your reading, as long as you can read the great green book, of which every bud is a letter, and every tree a page?

Wonder if you will. You cannot wonder too much. That you might wonder all your life long, God put you into this wondrous world, and gave you that faculty of wonder which He has not given to the brutes; which is at once the mother of sound science, and a pledge of immortality in a world more wondrous even than this. But wonder at the right thing, not at the wrong; at the real miracles and prodigies, not at the sham. Wonder not at the world of man. Waste not your admiration, interest, hope on it, its pretty toys, gay fashions, fine clothes, tawdry luxuries, silly amusements. Wonder at the works of God.

You will not, perhaps, take my advice yet. The world of man looks so pretty, that you will needs have your peep at it, and stare into its shop-windows; and if you can, go to a few of its stage-plays, and dance at a few of its balls. Ah—well. After a wild dream comes an uneasy wakening, and after too many sweet things, comes a sick headache. And one morning you will awake, I trust and pray, from the world of man to the world of God, and wonder where wonder is due, and worship where worship is due.

PEAT BEDS

PEAT BEDS.

CHAPTER I.

THE FENS, THEIR STORY.

FANCY may linger, without blame, over the shining meres, the golden reed-beds, the countless water-fowl, the strange and gaudy insects, the wild nature, the mystery, the majesty—for mystery and majesty there were—which haunted the deep fens for many a hundred years. Not long ago, it was grand enough, that black ugly place, when backed by Caistor Hauglands and Holm Wood, and the patches of the primæval forest; while dark-green alders, and pale-green reeds, stretched for miles round the broad lagoon, where the coot clanked, and the bittern boomed, and the sedge-bird, not content with its own sweet song, mocked the notes of all the birds around; while high overhead hung, motionless, hawk beyond hawk, buzzard beyond buzzard, kite beyond kite,

as far as eye could see. And terrified wild-fowl screaming,
piping, clacking, croaking, filling the air with the hoarse
rattle of their wings, while clear above all sounded the
wild whistle of the curlew, and the trumpet note of the
great white swan

They are all gone now. No longer do the ruffs trample
the sedge into a hard floor in their fighting-rings, while
the sober reeves stand round, admiring the tournament of
their lovers, gay with ears and tippets, no two of them
alike.   Gone are ruffs and reeves, spoonbills, bitterns,
avosets ; the very snipe, one hears, disdains to breed.
Gone, too, not only from Whittlesea, but from the whole
world, is that most exquisite of English butterflies—*Ly-
cæna dispar*—the great copper ; and many a curious insect
more.   But there is a more wonderful past over which we
must linger awhile and see what the old Fen was.

To do that rightly, one must go back, it seems to
me, to an age before all history ; an age which cannot be
measured by years or centuries ; an age shrouded in
mystery, and to be spoken of only in guesses.   To assert
anything positively concerning that age, or ages, would be
to show the rashness of ignorance.   "I think that I believe," "I have good reason to suspect," "I seem to see,"
are the strongest forms of speech which ought to be used
over a matter so vast, and as yet so little elaborated.

"I seem to see," then, an epoch after the impervious
bedding of the fens had been deposited ; after the chalk
had been laid on the top of them, at the bottom of some
ancient ocean ; the boulder-clay had been spread out in

the " age of ice " on top of all ; after the whole had been
upheaved out of the sea. In those ages—while the valleys
of the Cam, the Ouse, the Nene, the Welland, the Glen,
and the Witham were sawing themselves out—I " seem to
see " a time when the straits of Dover did not exist—a
time when a great part of the German Ocean was dry land.
Through it, into a great estuary between North Britain and
Norway, flowed together all the rivers of north-eastern
Europe—Elbe, Weser, Rhine, Scheldt, Seine, Thames, and
all the rivers of east England, as far north as the Humber.

And if a reason be required for so daring a theory—a
sufficient one may be found in one look over a bridge, in
any river of the east of England. There we see various
species of Cyprinidæ, " rough " or " white " fish—roach,
dace, chub, bream, and so forth, and with them their
natural attendant and devourer, the pike.

Now these fish belong almost exclusively to the same
system of rivers—those of north-east Europe. They
attain their highest development in the great lakes of
Sweden.

Again, there lingers in the Cam, and a few other rivers
of north-eastern Europe, that curious fish the eel-pout
or "burbot." Now he is utterly distinct from any
other fresh-water fish of Europe. His nearest ally is the
ling ; a deep-sea fish, even as his ancestors have been.
Originally a deep-sea form, he has found his way up the
rivers, even to Cambridge, and there remains. The rivers
by which he came up, the land through which he must
have passed, ages and ages since, have been all swept away.

The next proof is the presence in the Fens—where he is now probably extinct—and in certain spots in East Anglia, which this writer will take care not to mention, of that exquisite little bird the "Bearded Tit." Tit, he is none; rather, it is said, a finch, but connected with no other English bird. His central home is in the marshes of Russia and Prussia; his food the molluscs which swarm among the reed-beds where he builds; and feeding on those from reed-bed to reed-bed, all across what was once the German Ocean, has come the beautiful little bird with long tail, orange tawny plumage, and black moustache which might have been seen forty years ago in hundreds on every reed-round of the Fen.

One more proof—for it is the heaping up of facts, each minute by itself, which issues often in a sound and great result. In draining Wreatham Mere, in Norfolk, not so very far from the Fens, in the year 1856, there were found embedded in the peat moss, remains of an ancient lake-dwelling, supported on piles. A dwelling like those which have lately attracted so much notice in the lakes of Switzerland; like those which the Dyaks make about the ports and rivers of Borneo; dwellings invented, it seems to me, to enable the inhabitants to escape not only wild beasts, but malaria and night frosts; and, perched above the cold and poisonous fogs, to sleep, if not high and dry, at least high and healthy.

In this district had been previously found, in the peat, two shells of the fresh-water tortoise, till then unknown in England.

LAKE DWELLINGS.

These little animals, which may be seen in hundreds in the meres of eastern Europe, sunning their backs on fallen logs, and diving into the water at the sound of a footstep, are eaten largely in continental capitals. They may be bought at Paris, at fashionable restaurants. Thither they may have been sent from Vienna or Berlin; for in north France, Holland, and north-west Germany they are unknown. A few specimens have been found buried in peat in Sweden and Denmark; and there is a tale of a live one having been found in the extreme south part of Sweden, some twenty years ago. Into Sweden, then, as into England, the little fresh-water tortoise had wandered, as to an extreme limit, beyond which the change of climate, and probably of food, killed him off.

But the emys which came to the Wreatham bog must have had a long journey; and a journey by *fresh* water too. Down Elbe or Weser he must have floated, ice-packed, or swept away by flood, till somewhere off the Doggerbank, in that great network of rivers which is now open sea, he or his descendants turned up Ouse and Little Ouse, till they found a mere like their old Prussian one, and there founded a tiny colony for a few generations, till they were eaten up by the savages of the table dwelling; or died out —as many a human family has died out—because they found the world too hard.

It was thus necessary, in order to account for the presence of some of the common animals of the fen, to go back to an epoch of immense remoteness.

And how was that great lowland swept away? Who

can tell? Probably by no violent convulsion. Slow
upheavals, slow depressions, there may have been—indeed
must have been—as the sunken fir-forests of Brancaster,
and the raised beach of Hunstanton, on the extreme north-
east corner of the Wash, testify to this day. But the
main agent of destruction has been, doubtless, that same
ever-gnawing sea-wash which devours still the soft strata

SEA-WASH.

of the whole east coast of England, as far as Flam-
borough Head; and that great scavenger, the tide-
wave, which sweeps the fallen rubbish out to sea
twice in every twenty-four hours. Wave and tide by
sea, rain and river by land; these are God's mighty
mills in which He makes the old world new. And
as Longfellow says of moral things, so may we of
physical:

"Though the mills of God grind slowly, yet they grind exceed-
     ing small.
Though He sit, and wait with patience, with exactness grinds
     He all."

The lighter and more soluble particles, during tha⸱

ANCIENT FEN-LANDSCAPE.

slow but vast destruction which is going on still to this
day, have been carried far out to sea, and deposited as
ooze. The heavier and coarser have been left along the
shores, as the gravels which fill the old estuaries of the
east of England.

From these gravels we can judge of the larger animals which dwelt in that old world. About these lost lowlands wandered herds of the woolly mammoth, *Elephas primigenius*, whose bones are common in certain Cambridge gravels, whose teeth are brought up by dredgers, far out in the German Ocean, off certain parts of the Norfolk coast. With them wandered the woolly rhinoceros (*R. tichorhinus*), the hippopotamus, the lion—not (according to some) to be distinguished from the recent lion of Africa—the hyæna, the bear, the horse, the reindeer, and the musk ox; the great Irish elk, whose vast horns are so well known in every museum of northern Europe; and that mighty ox, the *Bos primigenius*, which still lingered on the continent in Cæsar's time, as the urus, in magnitude less only than the elephant,—and not to be confounded with the bison, a relation of, if not identical with, the buffalo of North America, which still lingers, carefully preserved by the Czar, in the forests of Lithuania.

But did all these animals exist at the same time? It is difficult to say. The study of the different gravels is most intricate—almost a special science in itself—in which but two or three men are adepts.

It is hard, at first sight, to believe that the hippopotamus could have been the neighbour of the Arctic reindeer and musk ox; but that the woolly mammoth not only may have been such, but was such, there can be no doubt. His remains, imbedded in ice at the mouth of the great Siberian rivers, with the wool, skin, and flesh (in some cases) still remaining on the bones, prove him to have been fitted for a cold climate, and to have browsed upon the scanty shrubs of Northern Asia. But these huge mammals, now confined to hotter countries, once inhabited a colder region, or at least wandered northwards in whole herds in summer, to escape insects, and find fresh food and, above all, water.

The same is the case with the lion, and other huge beasts of prey.

The tiger of Hindostan ranges, at least in summer, across the snows of the Himalaya, and throughout China. Even at the river Amoor, where the winters are as severe as at St. Petersburg, the tiger is an ordinary resident at all seasons. The lion was, undoubtedly, an inhabitant of Thrace as late as the expedition of Xerxes, whose camels they attacked; and the "Nemæan lion," and the other lions which stand out in Grecian myth, as having been killed by Hercules and the heroes, may have been the last remaining specimens of that *Felis spelæa* (undistinguishable, according to some, from the African lion), whose bones are found in the gravels and the caverns of these isles.

PEAT BEDS.

CHAPTER II.

THE FENS, THEIR AGE.

A ND how long ago were those days of mammoths and reindeer, lions and hyænas? We must talk not of days, but of ages; we know nothing of days or years.

Indeed, we are utterly at a loss for a definite answer. The interval of time may have been very great. But we have no scale on which to measure it.

Let us suppose, then, the era of " gravels " past; the valleys which open into the fen sawn out by rivers to about their present depth; and what life was it which peopled the fens?

The all-devouring, yet all-preserving peat-moss contains our answer. Gradually it swallowed up the stately forests of fir and oak, ash and poplar, hazel and yew, which once grew on that rank land, and the beasts which in savage freedom roamed amongst them.

The gigantic ox, *Bos primigenius*, was still there, but with him appeared another ox, smaller and with shorter horns, *Bos longifrons;* which is held to be the ancestor of our own domestic short-horns, and of the wild cattle still preserved at Chillingham and at Cadzow. The rein-deer had disappeared, almost or altogether.

The red deer, of a size beside which the largest Scotch stag is puny, and even the great Carpathian stag inferior, abounded; so did the roe; so did the goat, which one is accustomed to look on as a mountain animal. In the Woodwardian Museum there is a portion of a skull of an ibex—probably *Capra sibirica*—which was found in the drift-gravel at Fulbourne. Wild sheep were unknown.

The horse occurs in the peat; but whether wild or tame, who can tell? Horses enough have been mired and drowned since the Romans set foot on this island, to account for the presence of horses' skulls, without the hypothesis of wild herds, such as doubtless existed in the gravel times.

The wolf, of course, was common; wild cat, marten, badger, and otter, all would expect; but not so the beaver, which nevertheless is abundant in the peat; and damage enough the busy fellows must have done, cutting trees, damming streams, flooding marshes, and like selfish specu-lators in all ages, sacrificing freely the public interest to their own.

Here and there are found the skulls of bears, in one

case that of a polar bear, ice-drifted; and one of a walrus, probably washed in dead after a storm.

Beautiful, after their kind, were these fen-isles, in the eyes of the monks who were the first settlers in the wilderness. "It represents," says one, "a very Paradise, for that in pleasure and delight it resembles heaven itself. These marshes abound in trees, whose length without a knot doth emulate the stars. The plain there is as level as the sea, which with green grass allures the eye, and so smooth that there is nought to hinder him who runs through it. Neither is therein any waste place; for in some parts are apple trees, in others vines, which are either spread on the ground or raised on poles. A mutual strife is there between nature and art; so that what one produces not, the other supplies. In it was," to quote roughly, "abundance of tame beasts and of wild stag, roe, and goat, in grove and marsh; martens, and ermines, and fitchets, which in hard winter were caught in snares or gins. But of the kind of fish and fowl which bred therein, what can I say? In the pools around are netted eels innumerable, great water wolves, and pickerel, perch, roach, burbot, lampreys, which the French called sea-serpents; smelts, too; and the royal fish, the sturgeon. But of the birds which haunt around, if you be not tired, as of the rest, we will expound. Innumerable geese, gulls, coots, divers, water-crows, herons, ducks, moulting time, I have seen hundreds taken at a time, by nets, springes, or birdlime," and so forth; till, as he assures William, the Frenchman may sit on

Haddenham field blockading Ely for seven years more,
" ere they will make one ploughman stop short in his
furrow, one hunter cease to set his nets, or one fowler to
deceive the birds with springe and snare."

ANCIENT DWELLING.

And yet there was another side to the picture. Man
lived hard in those days, under dark skies, in houses—
even the most luxurious of them—which we should think,
from draughts and darkness, unfit for felons' cells.

Hardly they lived; and easy were they pleased, and
thankful to God for the least gleam of sunshine, the least
patch of green, after the terrible and long winters of the
Middle Age. And ugly enough those winters must have
been, what with snow-storm and darkness, flood and ice,
ague and rheumatism; while through the dreary winter
nights the whistle of the wind and wild cries of the
water-fowl were translated into the howls of witches and
demons; and (as in St. Guthlac's case) the delirious
fancies of marsh fever made those fiends take hideous
shapes to the inner eye, and act fantastic horrors round
the old fen-man's bed of sedge.

There you may read, too, a detailed account. There
were creatures, at least so he says, who used to hale St.
Guthlac out of his hut, drag him through the bogs, carry
him aloft through frost and fire—"Develen and luther
gostes"—such as tormented in likewise St. Botolph
(from whom Botulfston = Boston, has its name), and
who were supposed to haunt the meres and fens, and to
have an especial fondness for old heathen barrows with
their fancied treasure-hoards; how they "filled the
house with their coming, and poured in on every side,
from above, and from beneath, and everywhere. They
were in countenance horrible, and they had great heads,
and a long neck, and a lean visage; they were filthy and
squalid in their beards, and they had rough ears, and
crooked 'nebs,' and fierce eyes, and foul mouths; and
their teeth were like horses' tusks; and their throats
were filled with flame, and they were grating in their

voice; they had crooked shanks, and knees big and great behind, and distorted toes, and cried hoarsely with their voices; and they came with such immoderate noise and immense horror, that he thought that all between heaven and earth resounded with their voices. . . . . And they tugged and led him out of the cot, and led him to the swart fen, and threw and sunk him in the muddy waters. After that they brought him into the wild places of the wilderness, among the thick beds of brambles, that all his body was torn. . . . . Then they took him and beat him with iron whips, and, after that, they brought him on their creaking wings between the cold regions of the air."

Does not all this sound—as I said just now—like a voice from another planet? It is all gone; and it was good and right that it should go when it had done its work.

THE END.